IN THE BEGINNING

IN THE BEGINNING

WORLD HISTORY FROM HUMAN EVOLUTION TO THE FIRST STATES

LAUREN RISTVET

Georgia State University

Boston   Burr Ridge, IL   Dubuque, IA   Madison, WI   New York
San Francisco   St. Louis   Bangkok   Bogotá   Caracas   Kuala Lumpur
Lisbon   London   Madrid   Mexico City   Milan   Montreal   New Delhi
Santiago   Seoul   Singapore   Sydney   Taipei   Toronto

Higher Education

IN THE BEGINNING: WORLD HISTORY FROM HUMAN EVOLUTION TO THE
FIRST STATES

Published by McGraw-Hill, a business unit of The McGraw-Hill Companies, Inc., 1221
Avenue of the Americas, New York, NY, 10020. Copyright © 2007 by The McGraw-Hill
Companies, Inc. All rights reserved. No part of this publication may be reproduced or
distributed in any form or by any means, or stored in a database or retrieval system,
without the prior written consent of The McGraw-Hill Companies, Inc., including, but
not limited to, in any network or other electronic storage or transmission, or broadcast
for distance learning.

Some ancillaries, including electronic and print components, may not be available to
customers outside the United States.

This book is printed on acid-free paper.

1 2 3 4 5 6 7 8 9 0 DOC/DOC 0 9 8 7 6

ISBN: 978-0-07-284803-8
MHID: 0-07-284803-0

Vice President and Editor-in-Chief: *Emily Barrosse*
Publisher: *Lyn Uhl*
Senior Sponsoring Editor: *Jon-David Hague*
Editorial Coordinator: *Sora Kim*
Marketing Manager: *Jennifer Reed*
Managing Editor: *Jean Dal Porto*
Project Manager: *Jean R. Starr*
Art Director: *Jeanne Schreiber*
Art Editor: *Ayelet Arbel*
Designer: *Srdjan Savanovic*
Photo Research Coordinator: *Natalia C. Peschiera*
Production Supervisor: *Janean A. Utley*
Composition: *10/13 Palatino, by Techbooks*
Printing: *R. R. Donnelley & Sons*

Credits: The credits section for this book begins on page C1 and is considered an
extension of the copyright page.

Library of Congress Cataloging-in-Publication Data
Ristvet, Lauren.
 In the beginning : world history from human evolution to the first states/
Lauren Ristvet. – 1/e.
  p. cm.
 Includes bibliographical references and index.
 ISBN-13: 978-0-07-284803-8 (pbk. : alk. paper)
 ISBN-10: 0-07-284803-0 (pbk. : alk. paper) 1. Prehistoric peoples. 2. Human
evolution. 3. Civilization, Ancient.
 I. Title.
  GN720.R57 2007
  930--dc22   2006033535

The Internet addresses listed in the text were accurate at the time of publication. The inclu-
sion of a Web site does not indicate an endorsement by the authors or McGraw-Hill, and
McGraw-Hill does not guarantee the accuracy of the information presented at these sites.

www.mhhe.com

▓ TABLE OF CONTENTS ▓

◖ NOTE FROM THE SERIES ◗
EDITORS

World History has come of age. No longer regarded as a task simply for amateurs or philosophers, it has become an integral part of the historical profession, and one of its most exciting and innovative fields of study. At the level of scholarship, a growing tide of books, articles, and conferences continues to enlarge our understanding of the many and intersecting journeys of humankind framed in global terms. At the level of teaching, more and more secondary schools as well as colleges and universities now offer, and sometimes require, World History of their students. One of the prominent features of the World History movement has been the unusually close association of its scholarly and its teaching wings. Teachers at all levels have participated with university based scholars in the development of this new field.

The McGraw-Hill series—Explorations in World History—operates at this intersection of scholarship and teaching. It seeks to convey the results of recent research in World History in a form wholly accessible to beginning students. It also provides a pedagogical alternative to or supplement for the large and inclusive core textbooks that are features of so many World History courses. Each volume in the series focuses briefly on a particular theme, set in a global and comparative context. And each of them is "open-ended," raising questions and drawing students into the larger issues which animate World History.

This book by Lauren Ristvet addresses the various "beginnings" of the human story—the slow emergence of humankind, the immensely long era of gathering and hunting societies, the several breakthroughs to agriculture, the development of pastoralism, and the first cities, states, and civilizations. Her account of these beginnings is informed by the most recent historical and archeological scholarship; it is thoroughly global and comparative, juxtaposing cases from both the eastern and western hemispheres; it highlights cross-cultural interactions and connections; and it is written in an evocative and enticing language altogether accessible to undergraduates. For teachers, students, and anyone else interested in the "ways we were" in the beginning, Professor Ristvet's book is a tour-de-force, a marvelous guide to the early stages of the human journey.

Robert Strayer
Kevin Reilly

◖ ACKNOWLEDGMENTS ◗

More people than I can possibly acknowledge here have helped me with this book. The series editors, Kevin Reilly and Robert Strayer, read and extensively commented on the manuscript, while Arlette McNeill and Jean Starr were helpful once the book was in production. David Christian reviewed the final manuscript and contributed several helpful references, as did eleven anonymous reviewers. Of other friends and colleagues, Sarah Parcak and Colleen Manassa supplied me with information about Egypt; Jason Nesbitt critiqued my discussion of Andean archaeology; and Eleanor Blue, Edward Castleton, Andrew Drabkin, Julie Elkner, and Athena Smith proofread various drafts. Students in my world history courses at Georgia State University read this book in manuscript and helped me correct difficult (and rewrite boring) passages. Finally, I never could have written this book without the example of three of my teachers—Gareth Jenkins, Harvey Weiss and Nicholas Postgate—who made my own first exploration of archaeology and early history endlessly fascinating.

 INTRODUCTION

OUTLINE OF THE INTRODUCTION

The Study of Early History
    Learning About the Past
    Looking at History

GETTING STARTED ON THE INTRODUCTION: When does history begin? Why and when do societies change and why do certain things stay the same? This chapter will provide an interpretive framework for the study of transition and present some of the main themes of the book.

LEARNING ABOUT THE PAST

When does history begin? Until recently, most European and American history books began with the Greeks. Chinese history books have traditionally begun with the Xia, the first dynasty to rule China in the early second millennium BC. Other historians and nonhistorians have often pointed to other events, such as the appearance of humankind, the birth of the gods, or the creation of the world. If asked to write a world history, an anthropologist might write about the past 6 million years since "hominids" (human ancestors) diverged from "hominoids" (humans, apes, and baboons). Biologists would probably place the starting line earlier, with the beginning of life, 3.5 billion years ago; whereas geologists might propose the date for the origin of the planet, 4.5 billion years ago. Astronomers might seek to push our story back to the origins of the universe, approximately 13.7 billion years ago. On the other hand, philosophers and theologians might dismiss this quest for origins, claiming that there can be no true beginnings. We will follow the anthropologist's version of history, beginning with the period between 6 and 8 million years ago when humans and chimpanzees diverged from a common ancestor.

Looking at History

Making sense of nearly 6 million years of human life can be difficult without some sort of evaluative framework. This book focuses on the twin themes of transition and maintenance, from both a regional and a global perspective. Essentially, this is a study of the most important transitions in human history: the origins of language, agriculture, cities, writing, and government. At the same time, much of history consists of recognizing what did not change from era to era and why: the study of maintenance or conservation. Eras during which little monumental change occurred often lasted for thousands of years. It is just as important to understand the absence of change as it is to understand change. Additionally, by examining societies prior to transition, we can ask better (and more interesting) questions about what changed and why.

TRANSITIONS

None of these transitions had to happen, at least not in the sense we usually imagine. People had lived successfully as hunters and gatherers for millions of years; they understood that plants came from seeds, but had never bothered to plant them before. So why did they decide to shift to plant cultivation, especially when life is often harder for farmers than it is for hunters and gatherers? Similarly, village life had been an established fact for millennia in the Middle East when large cities such as Uruk were founded. Cities are not necessarily "better" than villages; in fact, they can be murderous—killing their inhabitants through a combination of poor nutrition, decreased sanitation, and increased aggression. Why did farmers all over the world suddenly decide to produce more than they needed for themselves, and thus allow cities to be built? Additionally, we have to recognize that all transitions do not happen the same way: some are quick, some are slow, and most of them are not global. Native Australians never invented cities; this was a transition that they did, quite happily, without.

GLOBAL VS. REGIONAL APPROACHES

In evaluating early history, we must also strike a balance between general and specific approaches, or global and regional ones. Each of these transitions happened in different ways in different places, often due to specific regional conditions. Obviously, people in southwest Asia domesticated wheat and not corn because the latter existed only on another continent. Similarly, people in the New World domesticated few animals (just the dog, guinea pig, turkey, and llama) because there were few animals around that lent themselves to domestication (no wild sheep, goats, cows, pigs, chicken, horses, water buffalo, or camels). The theme of each chapter will reflect a global process, while the case studies will provide illustrations on a regional level.

CONVERGENCE AND DIVERGENCE

The concepts of convergence and divergence provide yet another framework for examining world history. Convergence describes a coming together toward greater similarity, whereas divergence describes a drawing apart toward greater difference. We can use these concepts to explore comparative questions: to what extent did various societies follow similar or different historical trajectories? And we can use them to explore what happens when societies come in contact with one another. Convergence often occurs following cultural encounters, while divergence may occur as cultures lose contact with one another.

Throughout human history these processes have worked together or separately to produce myriad inventions, societies, and ways of life. Each transition presented an opportunity for humans around the world. Sometimes they responded to these in divergent manners; the Upper Paleolithic creative explosion, for example, meant that distinct cultures, in terms of very different tools, artwork, and economic bases, arose around the world. Occasionally, however, human experiences converged; Mesoamericans, Mesopotamians, and the Chinese all invented the same tool, writing, to solve very different problems. World history is not a simple story of the progress of all of the world's cultures to one unified, glorious future. Rather, it encompasses the stories of a wide range of peoples who sometimes followed the same ends and sometimes did not.

In essence, this book strives to balance change and continuity, small regions and the entire world, and convergence and divergence, to tell a story that captures the major elements of early human existence. Such a story is not, and cannot be, definitive. New evidence emerges yearly, from excavations, work on the human genome, and new translations. Similarly, new perspectives on existing evidence arise, forcing us to consider the past in a new light. As a result, this book does not attempt to answer all of the questions that a study of early history can produce. Instead, it seeks to be open ended, to encourage further inquiries that might spur new and better understandings about our collective past.

CHAPTER ONE

THE FIRST SIX MILLION YEARS

OUTLINE

GETTING STARTED ON CHAPTER ONE: During more than 99 percent of human history, since the human lineage diverged from the chimpanzee lineage 6 million years ago until the invention of agriculture 13,000 years ago, people lived as hunters and gatherers in a very different world than exists today. How did environmental changes influence human evolution? Why did people begin walking upright? How, when, and why did human intelligence and culture appear? Why did humans invent tools, and why was this process so slow? When and how did people first leave Africa? What were the unintended consequences of their migrations? How do the remains at either Lake Mungo or Dolni Vestoniçe help us understand how foragers lived?

1

For the first 99 percent of human history, people lived as hunters and gatherers. They did not work for money, write (or read), live in cities (or even villages) or farm. This period before the invention of agriculture is called the Paleolithic and encompasses the time when humans were evolving both culturally and physically. It can be difficult to pinpoint both the beginning and the end of this period. It began somewhere between six and eight million years ago, when humans and chimpanzees diverged from a common ancestor. It began to end 13,000 years ago, when farming and village life were invented. In some places, such as Australia, the Paleolithic ended only within the past few centuries, when European colonists brought metal and farming to this continent. Across the world, later developments—such as agriculture, states, and writing—all emerged from trends in the Paleolithic. To understand how the world operates today, we must first understand how the events of the Paleolithic created the world we now live in.

Studying the Paleolithic is both uniquely difficult and uniquely exciting. Unlike other periods of history, when we basically know the outline of events, new evidence and new interpretations of this period surface constantly. Each new fossil may force us to rewrite or refine our previous theories regarding human evolution. In the past few years, anthropologists have unearthed several fragments of bone, claiming that each belonged to "our earliest hominid [human] ancestors."[1] In 2004, scientists published the discovery of a new species of hominids, *Homo floresiensis*. These tiny hominids lived in an Indonesian island until at least 18,000 years ago—long after we thought all other human species had become extinct.[2] Yet over the last couple of years, a fierce debate over this find has broken out, with some anthropologists arguing that *Homo floresiensis* were actually just "deformed modern humans." The fossils just keep getting older, and the story of "where we came from" becomes ever more complicated.

This chapter will begin by analyzing how changing climatic and environmental conditions shaped—and continue to shape—human history. The long-term story of the Paleolithic is the story of human evolution; we will analyze two major changes in our ancestors' bodies and minds: bipedalism and the origins of the mind. We will also explore how people lived before agriculture, gender roles in the Paleolithic workplace, and the origins of technology. We will conclude with a description of life during the Upper Paleolithic (the late Paleolithic, after the development of modern humans and thin stone blades about 60,000 years ago) in two places oceans apart: the Czech Republic and Australia.

CLIMATE AND ENVIRONMENT

One of the best ways to think about the environmental context of early human history is to imagine watching geologic time on fast-forward, with plates colliding and breaking apart, mountains rising and then crumbling, plants and animals appearing and mutating constantly.[3] Such an image might best illustrate the concept of impermanence, and the necessity of adaptation throughout human history. Prehistory can seem like a time of stasis, when nothing really changed, but the fluidity of the landscape counters such an assumption. The last four geologic epochs, the Miocene, Pliocene, Pleistocene, and Holocene, when humans evolved, have been characterized by intense climatic change.

THE MIOCENE TO THE PLIOCENE: SHAPING HUMAN ORIGINS

Studying the geologic past can be counterintuitive; it involves reconsidering accepted beliefs. When we think about climate change today, we worry about global warming, not global cooling. Yet, for the past 30 million years, the major problem for life has not been rising temperatures, but falling ones. Before that time, there were forests on Antarctica and above the Arctic Circle and glaciers did not exist. Thirty million years ago, however, the world started to cool. Falling temperatures trapped water in ice, causing less rain to fall. The cold made the ancient forests that had covered much of the world during the Mesozoic, the time of dinosaurs, and the early Cenozoic, our era, disappear. This cooling led to a drier climate, greater climatic variability, and fragmented environments. The favored habitats of the last 20 million years have been grasslands: savannas and prairies. They began developing during the Miocene, the geologic epoch lasting from 24–5 million years ago.[4] In Africa, wide swathes of rain forest gave way to open forests alternating with savannas. The spread of the savannas led to a rise in large herbivores; horses, wild cattle, zebras, giraffes, and antelopes evolved to take advantage of this new type of vegetation. Primates had always been woodland creatures, but the emergence of the savanna favored monkeys and apes that could leave the trees. Such factors inspired the evolution of baboons and hominids, both of which moved into the grasslands adjacent to their East African forested homelands.

THE MESSINIAN SALINITY CRISIS AND THE EMERGENCE OF AUSTRALOPITH-ECINES: Between 6.4 and 4.5 million years ago, the period when the human and chimpanzee lineages probably diverged, global cooling led

to the buildup of polar ice in Antarctica and the formation of glaciers in the Southern Hemisphere. Approximately 6.4 million years ago, enough water had been frozen in the Antarctic ice cap and glaciers that the level of the Mediterranean Sea fell, isolating it from the Atlantic. The sea dried up, leading to a series of widespread droughts in Europe, Asia, and Africa, called the Messinian Salinity Crisis. The "crisis" consisted of repeated periods when the Mediterranean dried up completely and then refilled, leading to a long episode of extreme climatic variability with alternating periods of drought and regular rainfall.[5] It is precisely at this time of dramatic environmental change that the first human ancestors, the australopithecines, arose.

THE PLIOCENE AND THE TURNOVER-PULSE HYPOTHESIS: The next period of climate change occurred approximately 2.5 million years ago, during the Pliocene (5–1.6 million years ago). Across the world, climatologists have uncovered signs of a drier, cooler world—for example, expansion of grassland in South America and windswept dust in central Europe and China.[6] The dramatic decline of global temperatures caused the spread of arid grasslands within the savanna patchwork of Africa that led to a transformation across species; many species died out while myriad others appeared at this time. Geologists call this theory the turnover-pulse hypothesis. Animals suited to life on the plains—wildebeests, gazelles, and grassland rodents—evolved, whereas some woodland creatures went extinct. For hominids, this period roughly corresponds to the development of several closely related species. Large-toothed hominids known as robust australopithecines appeared in Southern and Eastern Africa; this appearance paralleled the emergence of other large-toothed plains animals. At roughly the same time, the first members of our genus—a biological term that indicates a division above species—*Homo* (*Homo habilis*) appeared. *Homo habilis* was the first hominid to make stone tools.

THE ICE AGES AND THE HUMAN EXPERIENCE

THE PLEISTOCENE ICE AGES: The onset of the next geologic epoch, the Pleistocene (1.6 million–10,000 years ago) initiated another period of intense climate change. The Pleistocene has been characterized by 25 periods of glaciation, ice ages, and the warmer periods between them, interglacials. An example of an interglacial is our current epoch, the Holocene, which has lasted for only 10,000 years, and probably represents a brief warm period before the ice returns. During the Pleistocene, glaciers formed when slight drops in temperature increased snowfall and retarded melting. Currently, glaciers cover about 10 percent of the land surface of the earth, while during the colder periods of the Pleistocene, in contrast, they covered about 30 percent. In North America, these glaciers enveloped most of Canada and extended down to St. Louis, Missouri; in Europe they were found in Scotland, northern England, northern Germany, the Baltic states, and northwestern Russia. The movement of the glaciers formed many of the distinctive features in the landscape that we recognize today, such as the Finger Lakes of New York, the Lake District of England, and the valleys of the European Alps. Additionally, their formation meant that more water was locked in ice, causing sea levels to fall and altering the outlines of the continents as well as their relationships to islands and other continents. This fall in sea levels made the human settlement of Australia and the Americas possible during the late Pleistocene.

THE NEANDERTHAL ADAPTATION TO THE PLEISTOCENE: For humans living during episodes of glaciation, cold temperatures made survival in northern latitudes difficult. Yet it was during this period that many hominids left Africa for the harsh climates of Europe and Asia.[7] The intense cold encouraged various cultural and physical adaptations. The stocky, powerful skeleton of *Homo Neandertalensis*, better known as Neanderthals, a species of hominids that became extinct only 28,000 years ago, evolved under the pressure of icy conditions as a physical adaptation that enabled them to survive very cold weather. Yet humans also lived during interglacials and were forced to adapt to the shifting plant and animal life during these warmer periods.

HOLOCENE CLIMATE CHANGE: Although the past 10,000 years have witnessed much smaller shifts in temperature, these have nonetheless had a profound impact on human history. Following the end of the last ice age around 12,000 BCE a sudden, sharp drop in global temperatures, called the Younger Dryas event, changed the distribution of vegetation, stimulating experiments with agriculture. A worldwide drought lasted from 2200–1900 BCE and wreaked havoc on the early civilizations of Mesopotamia, Egypt, and Palestine. The Medieval Warm Period, which allowed for agriculture on Greenland, an island that is frozen today, is one of the most recent periods of climate change. These examples illustrate that natural climate change continues to occur. Human societies have had to confront periods of drought and intense cold as a matter of course. Climate is a major factor in the ability of all life forms to survive; humans are no exception.

HUMAN EVOLUTION

From about 7 million to 100,000 years ago, prior to the evolution of modern humans, various species of hominids flourished in Africa, Asia, and Europe. Some of these are our direct ancestors; some are not. To understand how our own characteristics developed, we must understand how these hominids evolved. For most of us, the two most important points in this 7 million year period are the beginning and the end. The two key goals of many anthropologists are finding the first human ancestor—who must have lived between 4 and 6 million years ago—and the first "anatomically modern human"—who probably lived about 200,000 years ago. Two important physical transitions characterize these stages. First, the earliest physical characteristic that separates human ancestors from chimpanzee ancestors is the ability to walk upright: bipedalism. Walking on two feet freed hominids' hands for a variety of tasks, such as tool use. Recent research indicates that bipedalism developed about 6 million years ago, making it twice as old as anthropologists had previously thought. Second, the development of our own species, *Homo sapiens*, probably coincides with the appearance of larger brains, creative minds, language, and culture. At different times, researchers have believed that one or another of these transitions holds the answer to the question, "what makes us human?" Neither of these developments is straightforward, and our theories about them continue to change as new evidence emerges.

A QUICK GUIDE TO EVOLUTION

Before discussing two of the great problems in human evolution—bipedalism and the origins of human intelligence—it may be useful to summarize the basic ideas behind evolution. Charles Darwin explained evolution in *The Origin of Species* using two foundational concepts: variance and inheritance. No members of a species—or even members of a tribe—are identical. Some are shorter, some have longer arms, some are smarter, and so on; this is variance. Inheritance simply means that parents pass on their characteristics to their children. If two pigeons with extra-large eyes have offspring, it is likely that their offspring will also have extra-large eyes. In any population there is always competition over limited resources such as food and mates. Such competition, often described as natural selection, is important to two of the ways that certain traits spread. First, individuals who can better feed themselves (and stay out of the way of predators) are more likely to survive to breed and thus pass on their traits. Second, individuals who possess traits that are considered sexually attractive by their potential mates are likely to have more sex and thus

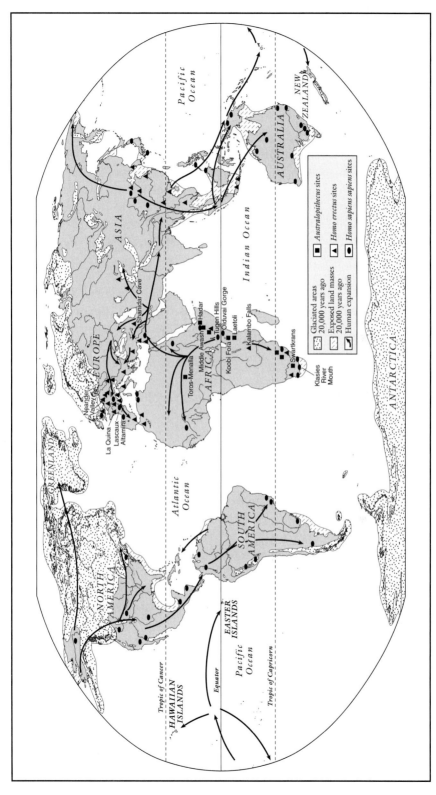

Global spread of hominids and Homo sapiens. For an interactive version of this map, go to www.mhhe.com/bentley3ch1maps.

more children than unattractive individuals. Over time, certain traits become dominant in a population, distinguishing it from other populations.

LEARNING TO WALK: THE ORIGINS OF BIPEDALISM

Who were our first ancestors? Although our very first ancestor, like that of all life-forms, was no doubt a single-celled organism, our first hominid ancestor was the first human-like primate who evolved into us, but not into other primates. In 2002, a team of French and Chadian anthropologists, working in Toros-Menalla, Chad (Figure 1, p. 28), uncovered a 7-million-year-old fossil, which they dubbed Toumai (*Sahelanthropus tahadensis*), a name taken from the local Goran language. These anthropologists think that *Toumai* closely resembles the common ancestor of chimpanzees and humans. Although he probably still mostly swung from trees like chimps, his teeth were more humanlike than chimpanzeelike.[8] Yet during the past year, a tense debate has raged in different scientific journals about whether Toumai is a hominid, a human ancestor, or a hominoid—a nonhuman primate—perhaps ancestral to the gorilla. Many anthropologists refuse to give Toumai proto-human status because there is no evidence that members of that species walked upright. In the past 40 years, the litmus test for human ancestors has often been whether they walked on two legs like us, not on four like other primates.

In the 1970s, anthropologists made two discoveries that received international attention and shaped the way we think about human evolution. In 1974, anthropologists found 40 percent of the skeleton of an *Australopithecus afarensis* in an excavation (usually they find only scattered teeth and disconnected bone fragments), allowing them to reconstruct how the entire skeleton of this small (one meter, or three and a half feet tall) woman once looked. Her femur, pelvis, and even vertebrae proved that this 3.18-million-year-old woman with an apelike face had walked upright. That night, in their campsite in Hadar, Ethiopia, Tim White and Donald Johanson played the Beatles' *Lucy in the Sky with Diamonds* to celebrate their discovery and playfully dubbed the skeleton Lucy. Two years later, Mary Leakey, working in Laetoli, Tanzania, discovered hominid footprints preserved in volcanic ash. The footprints had been left by two adult australopithecines, like Lucy, who 3.75 million years ago walked hand in hand, followed by a child over a dusty path to a water hole. These footsteps provide definite evidence of bipedalism; they are within the normal human range and resemble those made by people who have never worn shoes. These two discoveries received massive international attention; there are few natural history museums that do not have models of Lucy and the famous footprints on display.

So why should we care if our ancestors 3.5 million years ago walked upright or swung from trees? Bipedalism matters because it is the first physical transformation that we can trace in early hominids that makes them like humans and different from other primates. Walking upright allowed early hominids to avoid heat prostration because it exposed less of their body surface to the direct sun. According to some paleoanthropologists, this led to better temperature regulation of blood circulation, allowing brains to grow larger. Bipedalism let hominids see for long distances in an open environment, travel farther, and keep their hands free to carry objects like tools, food, and babies. Once arms and hands were no longer needed for locomotion, early humans could frighten predators by waving their arms or throwing stones. The very first accounts of bipedalism proposed that it evolved to allow for tool use and more efficient hunting. We now know that stone tools were not invented until at least 2 million years after bipedalism arose and that big-game hunting did not begin until at least 4 million years later. Ironically, bipedalism did not evolve to take advantage of what we see as its biggest benefit, but was the result of different evolutionary constraints.

How did bipedalism emerge? Why did human ancestors develop their characteristic gait when all other apes were happy to swing from trees, and occasionally walk on their knuckles? Until 10 years ago, explanations of bipedalism assumed that it reflected a move from the forests of East Africa into the savanna, as savanna habitats expanded due to climatic pressures. Australopithecines like Lucy and the footprint family lived in the African savanna, a wide grassland with few trees, as did most human ancestors beginning about 4 million years ago. Because other great apes, like chimpanzees and gorillas, live in forests, many anthropologists thought that adaptations for savanna living were what made humans different from other apes. We now know that this is not true. In 1994, a few fossils of *Ardepithecus ramidus*, an early human ancestor (dating to 4.5 mya), were unearthed in Ethiopia. These hominids probably walked upright; nonetheless, the fossils were found with fossil wood, seeds, and the bones of other forest-dwelling creatures, such as monkeys. This suggests that *Ardepithecus ramidus* lived in forests just like apes do today. Indeed, the recent discovery of several more fossils of early hominids who lived in woodland habitats has discredited the long-popular hypothesis that hominids became bipedal to adapt to life in the grasslands.

More very early bipedal hominids from forested environments have emerged, including the two new contenders for the "oldest human ancestor": the 6-million-year-old *Orrorin tugenensis,* also known as "millennium man," and the 5.8-million-year-old *Ardepithecus ramidus kadabba.*

Other paleoanthropologists have argued that later australopithecines also preferred to sleep and live in the safety of forests rather than in the savanna. Not even the famous Lucy is immune from these revisions; a new analysis of her skeleton suggests that her long arms let her swing from branches and that she slept in trees, much like modern chimpanzees.

Since these discoveries, discussion of the origins of bipedalism has flourished. Some researchers are reluctant to give up the savanna hypothesis and argue that bipedalism developed to allow early hominids to cope with a new environment of dispersed forest and grassland that developed from 10 to 5 million years ago. The creatures that would become bipedal were originally tree-feeding apes, forced to exploit many separate forested areas in this new environment. Initially, the open environments that these apes had to cross were useless stretches of empty grassland. They began to walk because it was the least awkward way to move around on the ground, and it let them get to their beloved forests faster. Other scholars have tried to link bipedalism and gender roles, suggesting that it arose to permit males to carry food back to the place where females lived. Still others have looked at primate behavior and proposed that bipedalism arose to permit males to display their penises more effectively to potential mates and thus increase their chances of finding a sexual partner.

All of these and other theories have their advocates and their critics. But no matter what purpose it originally fulfilled, bipedalism quickly became generally useful as a bevy of early bipedal hominids ranged over East and South Africa beginning about 4 million years ago. Bipedalism was an advance that stimulated evolutionary divergence, resulting in several species of hominids. Each of these species lived successfully for hundreds of thousands of years, eating different food, living in slightly different environments, and walking upright. Yet despite this proliferation of species, bipedalism does not seem to have been "the great leap forward" that early scholars hypothesized. Instead, the behavior of early bipedal hominids was probably closer to modern apes than to modern humans. Human intelligence and cultural diversity emerged much later.

LEARNING TO LEARN: THE ORIGINS OF INTELLIGENCE

In general, when faced with defining uniquely human characteristics, philosophers, anthropologists, and humanists stress our powers of reason, our self-awareness, and our ability to believe in abstract concepts. Human brains are nearly four times the size of chimpanzee brains. An increase in brain size was certainly one of the most important developments in making us human, but size alone cannot explain why humans developed such characteristics as language and art. Elephants and whales have huge brains, but clearly they have yet to design the equivalent of the Sistine Chapel. Neanderthals had larger brains than we do on average and yet were unable to innovate, relying on the same stone tools for thousands of years. In terms of understanding long-term human history, the development of the brain provides only the outline of the story.

The study of the origins of human intelligence has been controversial because it poses a fundamental philosophical question: what is the relationship between the brain (a physical organ) and the mind (an intellectual capacity for abstract thought and creativity)? Some scholars see the mind as an intrinsic part of the brain, stressing that a modern brain means a modern mind. Others view mind and brain as entirely separate and suggest that minds evolved well after our brains were essentially modern. This division also reflects inconsistencies between archaeological and biological evidence. Data for artistic expression, long-distance trade, and ocean voyages, which archaeologists believe coincide with the development of modern intelligence, do not neatly correspond to the emergence of anatomically modern human beings, *Homo sapiens sapiens*. Instead, until recently, it seemed as though all these innovations emerged sometime between 50,000 and 30,000 years ago, during the Upper Paleolithic Revolution. Archaeologists who study this period emphasize the suddenness of the transition by terming it a revolution. Although a transition that lasted 20,000 years may not seem sudden to us, it is very different from the nearly 6 million years of human history that preceded it, when the same tool designs were used for over a million years. These archaeologists believe that all these innovations relate to the development of the modern mind about 40,000 years ago.

So why is that a problem? Why don't all scientists accept the 40,000 BP (before present) date for the evolution of the mind? The problem is that most genetic and fossil evidence points to a much earlier evolution of anatomically modern humans, suggesting that the common ancestors of everyone alive today lived some 100,000 to 200,000 years ago in Africa. The date is far from definite, because of the difficulties of dating genetic changes. Mitochondrial DNA analysis suggests that all humans alive today are related to one woman who lived between 100,000 and 250,000 years ago in Africa, often playfully referred to as the African Eve,[9] while Y-chromosome analysis shows that all men alive today are the descendants of an African Adam—who probably lived about 70,000 years ago. For years, no fossil evidence dating to the crucial period between 100–200,000 years ago had been discovered in Africa. Luckily in 2002, anthropologists found three skulls in Ethiopia that are between 160,000 and 154,000 years old.[10] The same scientists also reanalyzed two other skeletons from Ethiopia, found in the 1960s, and proved that they are 195,000 years old. These skeletons represent the oldest evidence for our species.[11] Although some scientists continue to argue that anatomically modern humans arose in various regions of the world—from earlier hominid populations—this seems more and more unlikely.

But if our ancestors 200,000 years ago had modern bodies and brains, why did they wait so long to use them? It is difficult to argue that all humans 40,000 years ago, 60,000 years after they had first appeared on the scene, suddenly woke up one morning with modern minds. All explanations of what produced this shocking change among everyone alive everywhere, instantaneously, rest on untested mechanisms and are thus necessarily unscientific and illogical. Yet, this is the argument that the revolutionists, who argue that the mind is essentially separate from the brain, must make. In contrast, the evolutionists have to account for those 145,000 lost years. Why, if everyone alive on earth was capable of endless inventiveness in terms of tools, religion, and art, did they wait so long to paint caves, carve figurines, and invent fishhooks? We could argue that our dating of these developments is off, or that we simply have not found the oldest cave art yet, but archaeologists distrust such arguments.

Some scholars, however, have argued that the gap between the archeological and biological evidence may not be so great after all. Sally McBrearty and Alison Brooks, two archaeologists who specialize in African prehistory, have shown that it is mostly false, at least with regard to Africa.[12] The problem is that most people interested in the origins of modern intelligence have begun with the evidence of European cave paintings and have ignored evidence from Africa that points to a much earlier date for the evolution of culture. Finely made bone tools from the Congo, South Africa, and Botswana all date to between 90 and 100,000 years ago, showing that humans had modern intelligence even before they immigrated to Europe. Similarly, archaeologists have found large quantities of pigments, especially red ochre, which is currently used in some parts of Africa for body decoration and art, dating back to the same period. A mass of beads and ornaments made from ostrich eggshells complete the evidence for the African origin of art, long before the European cave paintings were imagined. We may expect more such early discoveries from Africa to come to light showing modern human behavior in the first modern humans as archaeologists explore new sites across the continent; there are fewer excavations in all of Africa than in Dordogne province, France. Further evidence from genetics may also be expected; a recent study of Y-chromosome DNA suggests that East Asian populations share a common ancestor between 35,000 and 89,000 years ago, which might date their immigration to Asia. Such evidence suggests that anatomically modern humans left Africa about 60,000 years ago.[13] Obviously, modern humans could not leave signs of their dizzying creativity in places that they had not yet reached. The appearance of art and culture around the world at slightly different times probably reflects the different dates for the spread of modern humans around the world, not different populations waking up to the fact that they had modern intelligence and could use it. All of this helps narrow the gap between biology and archaeology.

Although these data might indicate where and when our minds evolved for the first time, they do not address the question why. Three groups of researchers have used slightly different aspects of evolutionary theory to explain the development of human intelligence. One group believes that the human brain evolved because of the environmental stresses of early life. The second group believes intelligence evolved to make hominids more attractive to potential suitors. The third group believes that it evolved to deal with the social stresses of group living.

Hominids were originally imagined as incredibly weak creatures, saddled with useless fingernails rather than sharp claws, and miniscule teeth instead of useful fangs. Historians argued that they managed to survive only because they could outsmart their predators. Recently, researchers have modified this theory to argue that climatic instability encouraged the development of human intelligence.[14] Organisms have two ways to adapt to shifting climatic regimes: either they can seek out other environments that resemble the one they have lost, a process called habitat-tracking, or they can develop great flexibility and eke out a living in a wide range of habitats. Humans may have pursued both strategies and, in the process, extended their geographic range until it became global. In effect, early humans were great generalists; their large brains allowed them to quickly adapt to new environments following quick bursts of cooling or warming.

The second theory relies on sexual selection as the prime mover in human evolution. It argues that early humans evolved more intelligence to impress potential mates with their suave seduction techniques.[15] Human intelligence did not develop because it was intrinsically useful. Instead, our large, complex brains are the human version of the peacock's tail, which developed to impress peahens, increasing the peacock's mating success. Hominids chose to mate with smart hominids, resulting in smarter children. Language, art, and morality also may have evolved to impress potential mates.

The third theory relies on social selection, an extension of Darwinian theory to explain the evolution of human intelligence. Richard Byrne, an ethologist who studies chimpanzee and baboon behavior, argues that social situations have been the major stimulant to the large human brain and intelligence. He has observed that other primates use their intelligence to compete in high-stakes social contests for popularity and social status, rather than to feed themselves. Among chimpanzees, leaders change as support bases grow and shrink. Low-status males can make or break the careers of rising alpha males by switching alliances.[16] The rivalry between Yeroen and Luit, two chimps that live in Burgers' Zoo, Arnhem, Holland, illustrates chimpanzee political struggles. Yeroen began as the dominant male, but Luit managed to gain supporters and usurp his role. He did this by secretly flattering the female chimpanzees to win them over to his side. When Yeroen was present, Luit ignored the females, but when he was absent, Luit carefully groomed them and played with their infants. With help from his female friends, Luit managed to beat Yeroen and become the highest-ranking chimpanzee.[17] This social manipulation has evolutionary consequences because social success means that a male chimp will be able to impregnate more females. Byrne argues that human intelligence similarly evolved for use in competitive social situations; our modern attempts to get ahead in social situations are classic primate behavior.

In the end, we are left with an enigma. We know, based on genetic and skeletal evidence, when humans indistinguishable from us evolved. Archaeological discoveries of figurines, complex tools, and distinctive artistic styles help us date the emergence of culture. But we do not know why these humans evolved the various cultural forms that we think of as uniquely human. Archaeologists can argue that burials containing grave goods reveal the first signs of religion, a belief in the afterlife. They can argue that the technical explosion in stone and bone tools of 60,000 years ago is similar to the intellectual advances that have led in recent times to writing, printing, and computers. Unfortunately, they cannot give us the complete story.

STONE AGE ECONOMICS

Before the invention of farming 13,000 years ago, all humans lived by foraging. They did not produce their own food; rather they relied on wild foods, gathering plants and small animals, scavenging dead animals, and hunting. There are many ways to live off the land. Foragers can specialize in a certain type of prey, living mostly off reindeer or salmon, for instance, or they can use everything that the environment offers. Hunter-gatherers can move very little, keeping to a main camp during most of the year, or move almost constantly. To understand how people survived for 6 million years, it is necessary to explore how foragers make their living. This means focusing on both men's work and women's work and exploring how and why people invented tools to make this work easier.

WHAT FORAGERS DO FOR A LIVING

Before we explore how early hunters and gatherers lived, it is necessary to recognize and challenge the stereotypes that have pervaded previous discussions of these societies. In general, views have swung between two extremes: the ideas of the ignoble and noble savage. Early re-creations of the lives of "cavemen" characterized them as unenlightened animal-men that lived from day to day, constantly toiling to eke out a desperate existence. Such reconstructions of early human life depicted them as bloodthirsty, always ready to hunt or to battle. This view persisted until the 1960s, when fieldwork in southern Africa described the life of the San foragers as idyllic instead. The San had to work for only two days a week to gather enough food on which to live, allowing them to pursue other interests, such as chatting, singing, dancing, and playing games. A caloric analysis of their diet has suggested that the San were better fed than their nonforager neighbors, not on the brink of starvation.[18] Most surprisingly, these studies were conducted during one of the worst droughts in southern Africa in the twentieth century. The only inhabitants of the Kalahari who were not suffering from malnutrition during this period were the foragers. Farmers and herders had even taken to gathering mongongo nuts, the wild staple of the San diet. Other subsequent ethnographic work done in the 1960s added to this picture. Many of these studies were conducted by anthropologists to provide archaeologists with an ethnographic analogy to life in the Paleolithic. It was commonly assumed that the foraging way of life of modern hunter-gatherers was a "behavioral fossil" that had survived unchanged from the distant past. Work like this led one anthropologist to portray the Stone Age as "the original affluent society,"[19] a time when all people's needs and wants were easily satisfied and scarcity was unknown.

This interpretation also explained why foraging lasted so long as the only economic pursuit. Farming was not invented until 10,000 years ago because foragers were perfectly happy with their lives—not because they were inept. These views of Paleolithic life hearken back to earlier visions of the noble savage and a peaceful "state of nature," like Jean-Jacques Rousseau's depiction of the happy state of affairs before civilization or Karl Marx's idea of foraging life as an idyllic mode of production, "primitive communism." Nonetheless, this vision of foraging life may be as false as its opposite.

Beginning in the 1980s, anthropologists working with hunter-gatherers became concerned that most re-creations of early foraging life were based on untrue assumptions. Edwin Wilmsen, an anthropologist who works in the Kalahari, pointed out that the San do not live a blessed life, free from stress. Instead, they often experience conditions of near starvation. More importantly, they do not choose to live the way they do. Rather, English colonialism and postcolonial politics in Botswana have deprived them of any other opportunity. Nor do the San represent a population of Stone-Age people untouched by the modern world; their history has been inextricably tied together with other peoples' history for at least 4,000 years. Furthermore, this argument can be extended to almost any "isolated" primitive group in the world. No population has been truly isolated: the Australian aborigines, who are often assumed to be a pristine population, had contact with the natives of New Guinea, who were associated with Indian Ocean traders and thus the entire world. Similarly, the Yanomami of the Brazilian rain forest, trumpeted as a group with no outside contacts, were affected by the European invasion of South America like other native peoples.[20] If we want to use modern hunting-gathering societies to understand those in the distant past, we must do so carefully, emphasizing the different contexts of modern and ancient foragers, and consider modern foragers "as real people, not as a category," instead of as the "ahistorical residues of ancient foragers."[21]

Moreover, how hominids lived has changed radically over 6 million years. Archaeological data and comparisons with other primates suggest that australopithecines, *Homo habilis,* and *Homo erectus* all mostly lived from scavenging and gathering a limited range of plant food. Neanderthals added a hunting specialization to their economic pursuits. Modern foraging strategies began to emerge only with the appearance of our species 200,000 years ago. They developed tools that allowed them to use a wider variety of species than other hominids did (or other primates do). They were the first primates to exploit the sea, inventing fish hooks and probably fish nets (the first evidence of nets we have comes from 20,000 year-old art); they also invented tools (like mortars to produce flour) that let them use many previously inedible plants. These innovations meant that humans no longer needed to stick to a narrow range of environments but were free to colonize the world.

It is also important to emphasize how different foraging lifestyles can be when practiced by different peoples. Foraging groups that survive today are far from identical. The Eskimos of the Canadian Northwest territory have little in common with the Efe pygmies of Congo, for example. They both live off the land, but their environments, social practices, ideals, and problems are very different. Furthermore, we need to confront our image of these societies as simple or primitive. Foragers have developed complex mental maps of areas as large as 100 square kilometers, intricate kinship systems, multifaceted religious and philosophical notions, and a slew of traditions. The lives of modern foragers are no simpler than those of any other human society. Modern foragers have been too often dismissed as Stone-Age relics—primitive peoples that can be destroyed when they get in the way of governments or corporations. The generalizations that we draw must not be used to condemn them as "child-like," thereby allowing for mistreatment and their eventual eradication.

Here then are some generalizations about modern hunters and gatherers that are also true for Paleolithic foragers, or at least for the behavior of modern humans after 100,000 years ago. First, hunter-gatherers are mobile. Moving around a certain area allows them to exploit different resources at different times. They do not wander randomly, but instead migrate seasonally, keeping to a definite territory. Second, foragers exhibit a high degree of complex planning. They take care not to deplete the resources in any one area. They practice forms of population reduction such as dispersal or infanticide so that the carrying capacity of a particular area is never exceeded. Third, foragers live and work cooperatively. Unlike chimpanzees, who have a complex social hierarchy but feed individually, foraging is usually a communal activity for humans. Fourth, foragers often share the food that they accumulate. This is always true of meat, and is sometimes true of vegetables, which are often shared with those individuals who cannot collect them efficiently, such as pregnant women. Sharing, or reciprocal behavior, minimizes the risk to the group by giving each individual within it a chance of survival. This type of behavior makes hunting-gathering economies far more egalitarian than any other economic system.

This cooperative system probably evolved out of one similar to *primate feeding bands*, where bands live together but work separately. Adaptations such as tool invention allowed humans to become ever more efficient foragers. The development of stone tools and the development of other "tools," such as fire and boats, allowed early humans to exploit a greater range of environments. By 10,000 years ago, humans could be found on every continent, except Antarctica, living everywhere from forests to tundra to desert.

MAN THE HUNTER, WOMAN THE GATHERER?
DIVISION OF LABOR IN THE PALEOLITHIC

Before the 1960s, most anthropologists explained human evolution in terms of adaptations needed for more efficient hunting. In their portrait of the Paleolithic world, submissive women, saddled with pregnancies and small children, waited patiently for their fierce hunter husbands to supply them with carcasses. In this vision of history, man the hunter introduced all the important changes in prehistory. The development of bipedalism freed his hands for weapons. Increased brain size and language allowed his hunting expeditions to be planned more effectively. Finally, art and religion developed as a worldview that centered on the hunt, as seen in the paintings of wild horses, bulls, and reindeer at Lascaux (Figure 1). As a result, hunting, as a specifically male activity, was assumed to be important from the very beginning of human evolution. Raymond Dart, the paleoanthropologist who discovered the first australopithecine fossils, thought that the animal bones found mixed with these remains were evidence of the hunting prowess of early humans. Thus, he interpreted Swartkrans Cave, South Africa, in such a light, proposing that these Australopithecines had "[slaked] their ravenous thirst with the hot blood of victims and greedily devouring living writhing flesh."[22] Yet a careful reanalysis of the remains of the cave by geologists and archaeologists has rejected this bloodthirsty view of early humans. Leopard tooth marks found on a hominid skull provide evidence that rather than being the hunters, the individuals who found their way into Swartkrans were the hunted.

ORIGINS OF HUNTING: Other excavations have called into question any evidence of big-game hunting before the Middle Paleolithic and the evolution of Neanderthals and *Homo sapiens*, about 200,000 years ago. The patterning in the kills of elands (an antelope-like animal) at Klasies River Mouth, South Africa, an early *Homo sapiens* site dating from 100,000 years ago, indicates that hunters probably targeted these animals and hunted them by driving them into traps. Similarly, at La Quina, France, the piles of horse and reindeer bones found at the bottom of a cliff suggest that the Neanderthals drove herds of animals over this cliff, a technique used by

Native American hunters in more recent times. The lesson seems to be that regular big-game hunting is a difficult skill to master, one that arose late in human history.

THE SCAVENGING HYPOTHESIS: Chemical analysis of australopithecine and early *Homo* remains shows that they ate meat. This meat, however, mostly came from small-game hunting or scavenging. Chimpanzees and baboons do hunt and eat small animals; early humans probably inherited this primate ability as well. Meat from very large animals, represented at archaeological sites by bone remains, was probably gathered by opportunistic scavengers, not aggressive hunters. Although cut marks made by stone tools are found on the bones of large animals, they often overlie carnivore tooth marks. A comparison of the bone collection found at Olduvai Gorge, Tanzania, and modern scavenged bones collected on the Serengeti suggests that these ancient bone remains were the result of scavenging. Hominids probably stole carcasses that large carnivores had already killed and used tools to acquire the meat and marrow that carnivores and other more careless scavengers had left behind. They may have dragged these bones to caches—places where they stored stone or stone tools and then butchered and distributed the meat. Over time, this created the sites that archaeologists call home bases. Such sites may not have necessarily been the early hominids' homes, but rather spaces that the hominids used regularly for food processing. Unfortunately, as Donald Johanson dryly comments, "the emerging view of 'man the scavenger' did not have quite the heroic proportions of 'man the hunter.'"[23] All the same, it would appear to correspond better to the evidence in question.

THE IMPORTANCE OF GATHERING: But this picture of "man the scavenger" still portrays active men providing their passive women with food and thus stimulating the development of all those things that make us human, from *pair bonding* (monogamous couples) to the development of the brain and language. The archaeological record sometimes favors this interpretation. Bones last, while plant remains rarely do. We find evidence of plant consumption only at sites with unique preservation conditions, like Kalambo Falls, Zambia, a *Homo erectus* camp where leaves, nuts, seeds, fruit, and wooden tools were discovered that date to about 180,000 BP. So while archaeologists know that early hominids must have eaten some plants, they rarely emphasize this fact. However, fieldwork done with modern foragers has shown that plants are far more important than animals in the diet of hunter-gatherers. In tropical and temperate environments, women supply most of the calories consumed by modern foragers by gathering plants and small animals. It seems that focusing on big-game hunting is a poor strategy for feeding a band of hunter-gatherers. Indeed, given hunting's overall ineffectiveness and rarity among foragers today, it is absurd to suggest that male hunting or scavenging was the driving force behind human evolution. Theories about the dominance of "man the hunter" give insight not into the distant past, but into recent prejudices regarding the roles women should play in everyday life.

WOMAN THE HUNTER?   The sharp division between "man the hunter" and "woman the gatherer" has also been challenged by recent scholars. Male and female chimpanzees and other primates show great flexibility in survival activities. Jane Goodall has shown that female chimps participate in all aspects of daily social life and feed themselves and their offspring; they are even responsible for hunting about 35 percent of the animals chimps eat.[24] Other scholars have shown that primate females can take care of themselves and their children without male assistance. Ethnographic studies also amply document flexible behavior in forager females. Anthropologists know that the only true generalization that we can make about human behavior is that we cannot make any generalizations. It is unsurprising, then, that in some cultures men gather plant food and women hunt or scavenge. !Kung men of the Kalahari gather and eat mongongo nuts when they are hungry. Agta women of northeastern Luzon, in the Philippines, hunt and fish, often making their own equipment. Research on foragers has emphasized the flexibility of their behavior, which allows them to react to any opportunity. Although activities such as hunting and foraging certainly vary among individuals, this is due either to age or, more specifically, to the reproductive stage of females (late pregnancy, for example) rather than to essential gender differences. In short, it is important to realize that both men and women have been present throughout human evolution and human history, and that nothing is gained by ignoring or stereotyping the contributions of either sex.

FEMALE FIGURINES: Because men and women both played important roles in early economic life, it comes as no surprise that they similarly shared important positions in ritual life. Around 25,000 years ago, clay figurines of pregnant women appear across Eurasia. These "Venus" figurines (named after the Roman goddess of love and beauty), which represent the earliest human imagery, have been interpreted in several ways. Some feminists have argued that these statues depict powerful, fertile women and are the remnants of an ancient religion centered on goddess worship that flourished in a Paleolithic matriarchy (a society ruled by women). Other researchers believe that they represent Stone-Age pornography and that they "were made, touched, carved, and fondled by men" because "clearly no other group would have had such an interest in the female form."[25] Still other scholars believe they were self-portraits—the distortions reflect a woman looking down at her own body—or were toys for Paleolithic children. As in the debate about economic roles, this controversy seems to reflect modern prejudices regarding gender roles, rather than telling us what Paleolithic people thought about these roles. In any case, the very existence of these figurines shows that women participated in the creation of art (either as models and sculptors, or just as models) and suggests that they might have participated in religious life as well.

THE FIRST EDISONS: THE ORIGINS OF INVENTION

Thomas Carlyle, a nineteenth-century historian, once wrote: "man is a tool-using animal . . . without tools he is nothing, with tools he is all."[26] Many scholars have believed that tool use is a defining feature of humanity, which allowed human ancestors to transform their environment, instead of letting their environment transform them. Tools removed humans from certain evolutionary constraints. Rather than having to wait a hundred generations for natural selection to choose hairier humans who could better survive the Ice-Age climate of Europe, early humans could change that environment so it suited them, through the invention of clothing or the use of fire. Tool use allowed the peopling of the continents, the establishment of human society, and the creation of civilization. Or so the story sometimes goes. We know from research on animals that we are not the only species that uses tools. Chimpanzees use sticks, leaves, and rocks to procure termites, sap, and nut kernels. Even parrots use nutshells to obtain drinking water from inaccessible locations.

TOOL MANUFACTURE AND ABSTRACT THINKING: In that case, what makes human tool use so special? What is the essential difference between breaking a stone and using its edge as a knife, as *Homo habilis* probably did, and simply finding a sharp stone, as chimpanzees do today, to perform a task? Although any tool use requires a degree of planning, making tools, rather than using unmodified twigs, leaves, and rocks as other animals do, requires abstraction. One anthropologist tried to teach a bonobo named Kanzi to knap flint. Although Kanzi could produce sharp flakes, his mind did not have a grasp of the intuitive physics necessary to recognize how to work with the properties of the stone. It is perhaps this degree of abstract thinking that made *Homo habilis*, the first stone-tool makers, the first creative thinkers.

EARLY TOOL INDUSTRIES: But for most of human history, innovations in toolmaking were painfully slow. The Oldowan industry, the first toolset, consisting of flaked and smashed quartz riverside pebbles made into poorly formed choppers and sharp stone flakes, lasted for 1 million years with little change (from 2.6–1.5 million years ago). A second set of tools, characterized by heavy hand axes, appeared 1.5 million years ago and only disappeared 200,000 years ago, having lasted for 1.3 million years. This extreme conservatism provides a sharp contrast with modern human behavior, where new types of tools are invented constantly, without interruption. It's not as though these first manufactured tools, however primitive, were useless; they could have been used to work wood, cut reeds, and dismember animals. The anthropologist Louis Leakey once managed to skin a goat in only 20 minutes using Oldowan flakes. Nevertheless, the conservative behavior that characterizes early tool use emphasizes our own difference from these earlier hominids. Steven Mithen believes that *Homo erectus*, the makers of the handaxes, were very good at learning how to quickly survive in different environments using simple tools. Yet their knowledge of toolmaking, animal behavior, and environmental conditions remained in separate compartments of their mind; they were unable to integrate these different spheres of knowledge. As a result, they kept using the same tools for millions of years, because the architecture of their brains did not allow them to innovate.[27]

HARNESSING FIRE: Fire was another tool that early hominids used to expand their habitats and make foraging more efficient. *Homo erectus*, the maker of the famous handaxe, may also have invented fire. The earliest traces of human-managed fire come from Swartkrans, South Africa, about 1.6 mya, while another contemporary site in the Kenya Rift Valley also has stone artifacts and animal bones scattered around the baked clay of an early hearth. Fire would have allowed humans to live comfortably in the northern latitudes of Asia and Europe. It may be no coincidence that the data for the earliest use of fire and for the first settlements outside of Africa are contemporaneous. Fire also greatly expanded the range of foods that humans could eat—because certain indigestible (or poisonous) raw foods are excellent cooked. Similarly, humans who generally ate cooked meat could expect to live longer and be healthier than their contemporaries who enjoyed steak tartare, because cooking eliminates many toxins.

Indirectly, people could use fire both to hunt and to promote new vegetation growth for gathering. Yet recent research has questioned whether the hearths associated with these early hand axes really represent a human innovation, rather than simply the remains of a natural wildfire. Hearths do not become common at archaeological sites until the evolution of *Homo sapiens*. This suggests that the systematic use of fire was a late innovation in human history, associated with our own species.

It is difficult to date the invention of the other tools—clothing, shelter, and boats—that allowed early humans to populate the world. *Homo erectus* probably invented clothing, but no traces of this invention survive. Although all primates build "nests," either in trees or on the ground where they sleep at night, the first evidence of hominids building more substantial shelters are several 400,000-year-old huts from Terra Amata, France. The huts were erected in a line along the beach. They each contained a small living area with a hearth around which people cooked, knapped flint tools, and probably gossiped. Boats, the first constructed means of transportation, were probably invented last, not until our own species had evolved. The first evidence that we have for the use of boats is the settlement of Australia, about 60,000 years ago. The invention of stone tools, fire, shelter, and transportation meant that humans could survive all over the world.

"BE FRUITFUL, AND MULTIPLY, AND REPLENISH THE EARTH, AND SUBDUE IT"

Before the invention of farming, human populations were small and grew slowly. Demographers, scientists who study population growth, calculate that at the end of the Ice Age, the total population of hunter-gatherers was only 5 to 10 million. The highest possible population estimate is only 15 million people—lower than the population of Mexico City today. All this implies that population growth was very low—perhaps only .02 percent a year. In contrast, world population growth today is 1.2 percent a year, many times this rate. Nevertheless, estimates of population growth in prehistory are tentative, based as they are on environmental reconstructions, the behavior of modern hunter-gatherers, and sparse archaeological evidence. Growth rates probably varied by hominid species. It seems likely that after 100,000 years ago, *Homo sapiens* grew more quickly than their earlier ancestors, allowing them to migrate and colonize all the continents except Antarctica by the end of the Pleistocene.

Any species will grow until it can exploit all the habitats that are suited to it. Because culture allowed modern humans to adapt to almost all earthly environments, they quickly spread throughout the globe. This migration did not begin with modern humans. Australopithecines are found in both East and South Africa, separated by thousands of miles. *Homo erectus* skeletons are found in East Africa, Pakistan, and Georgia (in the Caucasus), and remains of their dwellings and stone tools are found in Spain, England, and Hungary. In fact, these prehistoric Marco Polos had even made it to China by 1.66 million years ago. By the time that *Homo sapiens* evolved, around 200,000 years ago, there were other species of humans living all over Africa, Europe, and Asia. Despite this, it seems obvious that the human migration was different from earlier hominid migrations—perhaps only because *Homo sapiens sapiens* managed to adapt their behavior to allow them to survive in an even larger range of environments. Boats and the ability of humans to adapt through culture to environments ranging between tropical forests and deserts allowed them to colonize Australia 60,000 years ago. Similarly, a cultural adaptation to extremely cold conditions allowed humans to settle Siberia and cross the land bridge between Asia and North America to settle the Americas by 15,000 years ago, if not earlier.

The process and effects of this early migration have been interpreted in two diametrically opposed fashions. Some scholars, especially in the nineteenth and early twentieth century, saw this early "peopling of the world" as a sign of both evolutionary and moral success. The quotation that heads this section, taken from Genesis, illustrates this idea—one that was current in both religious and scientific sectors until recently. Scholars and laypeople today are more inclined to see human evolutionary success—the ability to live anywhere and kill anything—as either immoral or, at the very least, morally ambiguous. Often, this quality, the ability to outcompete most other species, is seen as something that will lead (if it has not already led) to the destruction of the environment, and eventually the human race.

In the 1960s, scholars began to consider the implications of human colonization. Paul S. Martin, a paleontologist, pointed out that human migration into Australia and the Americas, continents that had never known any earlier hominid populations, was followed by a wave of extinctions of megafauna (large animals). The percentages involved are staggering: 73.3 percent of all animals over 100 pounds disappeared in North America; 79.6 percent of them died out in South America; and 86.4 percent of them became extinct in Australia. Although the naturalist Alfred Russell Wallace argued that a combination of man's activities and the end of the Ice Age caused these extinctions in the nineteenth century, prior to Martin's overkill hypothesis, most twentieth-century scientists had assumed these animals disappeared with the end of the Ice Age when the savanna habitats that they favored also disappeared. It is now clear, however, that the Holocene is no warmer than previous interglacials, which did not witness comparable extinctions. Similarly, the extinctions seem to have followed the establishment of human populations on these continents and not the end of the Ice Age. In Australia, the giant kangaroo, the marsupial lion, and a rhinolike mammal disappear about 30,000 years ago, when human settlers had occupied every habitat on this continent. Similarly, the extinctions of the mastodon, the camel, the saber-tooth tiger, and the musk oxen in North America coincide with an explosion of archaeological sites that appeared about 11,000 years ago, and which contain, somewhat sinisterly, a profusion of well-made flint tools called Clovis points, often found among the ribcages of these animals.

Martin argued that human hunters were able to decimate these species because they were an introduced predator. Mastodons had not evolved alongside people; therefore they had no defenses against them when human hunters suddenly appeared on the scene. Historical extinctions, like that of the famous dodo on Mauritius illustrate this process. When settlers reached Mauritius, these large flightless birds waddled up to them curiously, allowing the humans to club them to death leisurely. This process allowed Paleolithic hunters to kill these large, slow-breeding animals faster than they could reproduce. Over a few centuries, most of the large animal species in the Americas and Australia had been annihilated.[28]

Eurasia did not lose the same percentage of large animals as the Americas or Australia, although it also experienced megafauna extinctions. However, on this continent, the arrival of humans caused the extinction of the Neanderthals, a hominid species that had lived in Europe and Western Asia for 200,000 years. For a long time, Neanderthals were considered the ancestors of humans; however, archaeological finds of contemporaneous Neanderthal and human skeletons have shown that this was not true. A comparison of Neanderthal DNA and human DNA suggests that the two populations did not interbreed—meaning that the Neanderthals probably disappeared completely and do not live on in any descendants today.

How did humans decimate the Neanderthals? This was probably not done through any type of ancient warfare (it was certainly not an organized genocide), but instead had everything to do with demographics. Neanderthals and humans would have competed for the same resources.

In Europe, the human settlers used a wide range of new tools and resources, while the Neanderthals did not. Consequently, the humans were better fed and so had more children, who would have put more and more strain on resources that the Neanderthals used. Although there is some evidence that the Neanderthals tried to adopt the innovative techniques used by the modern humans (while many anatomically modern humans used techniques to make stone tools that the Neanderthals had pioneered), this adaptation seems to have been too little, too late. The last-known Neanderthal skeleton dates to 27,000 BP.

Prehistoric overkill and the disappearance of the Neanderthals are two "crimes" that recent anthropologists as well as environmentalists have laid at the feet of prehistoric people. It has been argued that the overkill hypothesis demolishes the myth of the noble savage, the belief that Paleolithic peoples, often popularly represented as the premodern equivalents of the indigenous Australians and Americans, lived in harmony with their environments. Similarly, the disappearance of the Neanderthals at human hands has been seen as yet another example of a species-wide tendency toward war and destructive violence. Although these arguments must be taken seriously, insofar as the evidence seems to support them, the particular historical context behind the interpretations also has to be emphasized. If the Victorians had known the same facts, they would have applauded, not derided, early man's prowess at hunting both animals as well as "degenerate humans" such as the Neanderthals. In short, we must never forget just how much interpretation of the past is shaped by the reigning prejudices of the present.

LIFE AT THE END OF THE ICE AGE

By 25,000 years ago, modern humans had established new cultures and diverse societies in Africa, Eurasia, and Australia. Agriculture would not be invented for another 15,000 years, but archaeological remains suggest that advanced hunting-gathering systems had evolved. Such societies used a broad range of resources. They ate fish and other marine life for the first time and hunted quite efficiently, often specializing in one species, such as reindeer in France. With the possible exception of the petite *Homo floresiensis* on the Indonesian island of Flores, the only surviving species of hominid 25,000 years ago was *Homo sapiens sapiens*, our own species. Asia, Africa, Europe, and Australia all boasted human populations. Fire had been tamed, art had been invented, and probably religion as well. With these inventions, as well as a proliferation of stone tools, the archaeological record changes and different artifacts are found in different areas. Culture, in the modern sense of a set of practices and a worldview understood only by a distinct group of people, and with it, diversity, had finally arrived. Stone tools and other materials in Africa looked very different from those in Australia. This variety reflected differences in ideals, ceremonies, and roles—none of which, unfortunately, can be observed directly now. To understand the lost world of the Stone Age, we shall consider life 25,000 years ago in two disparate places: the Czech Republic and Australia.

DOLNI VESTONIÇE, CZECH REPUBLIC

The remains of a settlement of mammoth hunters near Dolni Vestoniçe in the Czech Republic hints at the richness and complexity of daily life among these late Pleistocene hunter-gatherers. Dolni Vestoniçe's inhabitants were specialized mammoth hunters—able to exploit these mammals, who were one and a half times the size of a modern elephant. They used mammoth remains not only for meat, but also for fuel, construction, jewelry, and portable art. Perhaps mammoth parts were also useful in a Paleolithic trade network. Some of the shells found at Dolni Vestoniçe originated in the Mediterranean—suggesting that even 25,000 years ago, human settlements were not isolated, but were perhaps part of some larger system. Sculptures, pendants, necklaces, and headbands hint at a rich material life; people decorated themselves and their environment.

Paleolithic architects constructed huts from mammoth bones, earth, wooden posts, and mammoth hides. During the summer, people lived in ephemeral windbreaks, each of which housed one person or a small family, while during the winter, families huddled together in more substantial mammoth bone huts, grouped around a communal hearth. The hearth was probably at the heart of the Dolni Vestoniçe community, a place where people gathered and participated in various activities. The numerous huts at Dolni Vestoniçe imply almost permanent occupation, something that we generally expect from farmers—not foragers.

It seems likely that the inhabitants of Dolni Vestoniçe shared a rich symbolic life. To the west of the main settlement, a special round building partly built of limestone had been dug into the earth and contained the world's oldest ceramic figurines. The remains of 2,300 small clay figurines were found in an oven, covered by ashes and charred wood. They represent the only remnants of a long-forgotten ceremony, perhaps performed to encourage a successful hunt. These figurines were intentionally destroyed during firing; the potters moistened them so that they would

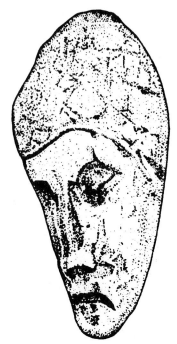
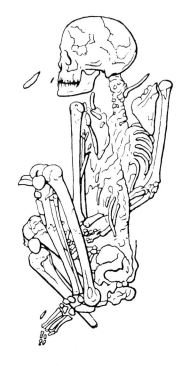

FIGURE 1
(*Left*) The head of a woman with an asymmetrical face, carved in ivory (4.6 cm high).
(*Right*) The buried skeleton of an elderly female from Dolni Vestoniçe. The bones of the left
side of the face revealed congenital nerve damage, probably resulting in an asymmetrical
facial expression. The woman was buried under two mammoth shoulder blades, and the
bones of an arctic fox were found next to her grave.

explode when placed in the fire. Some of these destroyed figurines were
ritually pierced with stone tools. Bird bone flutes, probably played during
this ceremony, lay scattered on the floor of the hut.

The people of Dolni Vestoniçe left us another clue to their symbolic
and moral life, in the form of two portraits in ivory. In 1936, an ivory
plaque with an incised human head was found. The face was lopsided;
the left-side drooped, as though its model had suffered a stroke. Twelve
years later, an ivory female head also portraying an asymmetrical face
was found in the summer hut (Figure 1). The following year, archaeologists found the grave of the woman who sat for these portraits. She had
been carefully laid to rest beneath two mammoth shoulder plates. An
analysis of her facial bones found evidence of congenital nerve damage to
the left side of her face, which had resulted in partial paralysis and a

lopsided face, like those preserved in ivory. The woman's body had been coated with red ochre, a pigment that has been used ritually around the world to symbolize blood and rebirth into the afterlife. Beside her lay the bones of an Arctic fox, perhaps a companion for her journey into death. The evidence of the woman's life and death reveals a society where people lived and worked together in a community that cared for them while they lived and after they died. Such a society does not seem so unalterably different from modern human societies.[29]

LAKE MUNGO, AUSTRALIA

At the same time that the mammoth hunters at Dolni Vestoniçe had settled down in their summer and winter huts, the original inhabitants of Sahul (greater Australia, including New Guinea and Tasmania) continued to be highly mobile. Their records include ephemeral sites, burials, and rock art. Australia has often been seen as a continent with little history, a place where life remained the same from first settlement until European contact. We now know that both economic strategies and cultural expressions did change over time. Despite this, Australia presents an interesting case of maintenance—not a place where change did not operate, but a place where the level of change remained low. Recently, new investigations into Australian prehistory have recovered evidence that Australian settlements are far older than previously suspected. It now seems likely that Australia was first settled around 60,000 years ago. Moreover, there is some evidence for an even earlier first settlement; new dates come in yearly, and are hotly disputed. For a continent whose culture changed little for millennia, Australia demonstrates the first evidence anywhere for sea-faring and painting.

Temporary campsites at Lake Mungo provide evidence of how Australians lived 25,000 years ago.[30] Today, the lake bed is dry and desiccated, but 25,000 years ago, the area would have been wetter. Remnants of ancient hearths—collections of charcoal, bird bones, eggshell, and mollusk shells—lie scattered among modern dunes. The early Australians responded to the scarcity of raw materials like stone (usually used in hearths) by inventing pottery containers for cooking. Excavations have also revealed large quantities of fishbones from cod, perch, and crayfish. A more recent find of a bone fishspear, dating back to 17,000 BP, suggests how the fish of Lake Mungo may have been harvested. Some of the animal bones around the lake came from animals that prefer arid habitats, indicating that the settlers at Lake Mungo traveled far in their pursuit of game.

Foragers spent a few weeks at Lake Mungo during the late winter and early spring. It may have been a stop on a seasonal route, one place among many where ancient Australians traveled. It was certainly one place where they buried their dead. Several burials were recovered from this site, including one of a woman who had been partially cremated—the oldest known cremation. Another body had been coated with red ochre, just like the woman found in the Czech Republic. These remains showed few signs of malnutrition or ill health. The evidence of the lives of these early Australians shows that far from being backward, they were just as creative as any other people at that time in the world.

CONCLUSION

The mammoth hunters at Dolni Vestoniçe and the seasonal travelers at Lake Mungo represent two very different foraging societies that flourished during the last Ice Age. The inhabitants of Dolni Vestoniçe lived most of the year at one site, where they could hunt herds of mammoths that visited the area regularly. Mammoth hunting formed the basis of their economy: they ate mammoth meat, used mammoth bones for their houses, and decorated mammoth ivory for rituals or for their own pleasure. The inhabitants of Lake Mungo took a different route. Rather than specializing in one particular animal, and living in one place year-round, they were generalists. They moved frequently and exploited all possible food sources, from the fish and plants found at Lake Mungo to desert animals found hundreds of kilometers away. Early Australian rock art suggests that these Australians had already developed a complex mythology. Some of the differences between the two sites can be explained by their different environments and climates. The extremely cold conditions in the Czech Republic 25,000 years ago encouraged the growth of large swathes of grassland, which, in turn, fed herds of animals such as mammoths. Lake Mungo, instead, was much drier and warmer, without huge herds, but with a wide variety of plant food and small-to medium-sized animals.

The remains found in the Czech Republic and Australia were left by our ancestors, individuals whose recognizably human behavior evolved during the last interglacial (130,000–80,000 years ago) and glacial (80,000–11,600 years ago). Both of these periods were more variable than the age we currently live in. Some years were very wet; some were very dry. The environment simply did not allow for the invention of agriculture until the Ice Age ended. Nevertheless, early anatomically modern humans

TIMELINE

Region	8 mya / 6 mya	5 mya	4 mya / 3 mya	2 mya / 1 mya	500,000 BP	250,000 BP	100,000 BP	50,000 BP / 10,000 BP
West Asia and North Africa				First human migration *Homo erectus*		Neanderthals	Migration of *Homo sapiens sapiens*	
East Asia				First human migration *Homo erectus*			Migration of *Homo sapiens sapiens*	
South Asia				First human migration *Homo erectus*			Migration of *Homo sapiens sapiens*	
Central Asia				First human migration *Homo erectus*			Migration of *Homo sapiens sapiens*	
Southeast Asia				First human migration *Homo erectus*			Migration of *Homo sapiens sapiens*	*Homo floresiensis*
Australia							Invention of boats; First settlement of Australia	Lake Mungo; Extinction of large mammals
Europe				First human migration *Homo erectus*		Neanderthals; First huts; Origins of big game hunting	Neanderthals; Migration of *Homo sapiens sapiens*	First figurines; Lascaux; Dolni Vestoniçe
West Africa				*Homo erectus*		Migration of *Homo sapiens sapiens*	First art	
East Africa	Divergence of human and chimpanzee lineages; First human ancestor? (Toumaï)	Development of bipedalism *Orrorin tugenensis, Ardepithecus ramidus kadabba*	Australopithecines; Invention of stone tools; Origins of language? (*Homo habilis*)	Invention of fire? *Homo erectus*		Emergence of *Homo sapiens sapiens*		
South Africa			*Homo habilis*			Migration of *Homo sapiens sapiens*		
North America								First settlement; Extinction of large mammals
Meso-america								First settlement; Extinction of large mammals
South America								First settlement; Extinction of large mammals

Miocene   *Pliocene (5.2 mya)*   *Holocene (13,000 BP)*   *Pleistocene (1.6 mya)* adapted to these conditions in a number of inventive ways. Their flexibility, demonstrated in the wide variety of environments that they lived in, foraging techniques that they used, and tools that they invented, is one of the defining characteristics of our species. Human history has been defined by this flexibility, this creativity, which allowed people to respond to a wide variety of opportunities by developing ever-changing cultures and societies.

CHAPTER TWO

THE ORIGINS OF AGRICULTURE

OUTLINE

GETTING STARTED ON CHAPTER TWO: The origins of agriculture and village life represent a major economic and social transformation, one that paved the way for the creation of states and civilizations. Why was agriculture invented during the Holocene and not before? Why did the beginning of agriculture follow different paths in different areas of the world? What were some of the unintended consequences of the origins of agriculture? What was the relationship of the origins of complex hunting-gathering societies to the origins of agriculture? Why did some societies never make the transition to agriculture? What was the relationship between the invention of the village, the nuclear family, property, and the roles of men and women in society? Did this vary geographically? Why?

The first word of Kurdish that my workmen taught me when I began to excavate in Northeastern Syria was *baran*, rain. During my first weeks there, I quickly picked up the terms for wheat, barley, sheep, goat, field, house, mother, father, and children. These simple words allowed me to talk casually to my workmen, all of whom were farmers. They now used John Deere combines to harvest their fields of wheat and barley, but they still grew the same crops that the first farmers did and lived in small agricultural villages made up of large families. Despite the enormous changes in technology, culture, and society since the origins of agriculture, some elements of life in a Western Asian agricultural village have remained the same.

Historians and social theorist have often stressed the importance of the origins of agriculture. Thomas Hobbes and Jean-Jacques Rousseau both saw it as the first step to village life, property ownership, and nuclear families. This view was refined by Friedrich Engels, working from notes left by Karl Marx; indeed, he called his book on the beginnings of agriculture—and ultimately, civilization—*The Origins of Family, Private Property, and the State*. If we think in this way, my simple Kurdish does not represent randomly chosen nouns, but instead the basics of an agricultural economy and society.

The establishment of agriculture across the globe during the past 10,000 years represents the most important economic revolution in human history. All recent inventions and economic developments rely upon this first step. The invention of agriculture allowed for the invention of computers, the railroad, and the first factories. Yet, the rise of agriculture did not proceed at the same pace, or use the same methods around the world. Instead, people everywhere were confronted with the same momentous event, the end of the Ice Age, but they responded to it differently. The trajectories that they followed help us to understand the Neolithic "revolution," which included the invention of agriculture and the village, regionally and globally. Agriculture was far more than a new way for humans to feed themselves; the changes it caused made it impossible for humans to revert back to a foraging lifestyle. In some places, agriculture was not invented or adopted by people until quite recently—often not until European colonialism. Instead, these peoples responded to the challenges of a postglacial world by developing new strategies to manage wild resources. To understand this revolution, we will consider both why people adopted agriculture and why they rejected it, and the economic, social, and cultural consequences of these decisions.

SETTING THE STAGE

Why did people invent agriculture? For years, philosophers and historians hardly even considered this question. Farming, taming the land through agriculture, was obviously superior to chasing after wild beasts or gathering grubs. Yet, agriculture is not actually easier than foraging—it requires more work and leaves its practitioners sickly and bored of a constant diet of gruel. In the recent past, few foragers, once exposed to agriculture, have adopted it without a fuss.

MUTUAL DEPENDENCE: What does it mean to invent agriculture anyway? Agriculture did not just "happen" because some Paleolithic Einstein figured out that if you planted a seed in the ground, carefully watered and tended it, it would grow into a beautiful flower (or a tasty bean plant). Animals were not simply domesticated because some softhearted child decided to make a pet out of a wounded goat. Foragers were not stupid; they had a very good knowledge of plant and animal habitats and they understood the scientific bases of agriculture for millennia.

Although early theorists considered agriculture one of the great human inventions, an innovation of an unknown genius, some scientists today see it as a less revolutionary and more natural process. In some ways, agriculture is not so different from the ways that foragers interact with plants. Agriculture may just represent a common type of mutual interaction between plants and animals, one which is not even limited to people. Consider the cultivator ant—an African ant that carefully prepares and tends fields of fungus. The "fruit" of this fungus, which the ants carefully harvest, forms the basis of their diet. Without the ants' loving care, the fungus would die; it cannot exist in the wild any more than domesticated corn can.[1] This perspective is repugnant to many people, who like to think that our creativity and mastery of the natural world, as represented by agriculture, is unique. It does suggest, however, that domestication may be a natural process, which is as beneficial to the plant as it is to us. After all, we have allowed wheat to conquer the world—greatly expanding its habitat. From the wheat's perspective, it is this plant that has exploited the silly humans.

WHY DID AGRICULTURE BEGIN DURING THE HOLOCENE?   Of course, this is an extreme way of looking at things. "Mutual dependence" aside, during the Pleistocene, people all over the world lived as hunters and gatherers, whereas during the Holocene, people all over the world domesticated plants and animals and began living as farmers. The major question about the origins of agriculture becomes not just "Why did people start farming?" but "Why did people start farming when and where they did, and not earlier or in other places?"

DEFINITION OF AGRICULTURE: Before we answer this question, we must consider what agriculture actually involves. The origins of agriculture mean the "domestication" of plants, which we can define as "growing a plant and thereby, consciously or unconsciously, causing it to change genetically from its wild ancestor in ways [that make] it more useful to human consumers."[2] The domestication of animals involves roughly the same process; it occurs only when domesticated animals become genetically different from their wild ancestors through human-controlled breeding. Agriculture is not just planting wild seeds; animal domestication is not just keeping wild animals in captivity (zoos are not in the business of domesticating cheetahs and chimpanzees!).

THE UPPER PLEISTOCENE: A recent theory claims that agriculture was impossible during the Pleistocene, but inevitable during the Holocene.[3] To understand this argument, we have to consider how the end of the Ice Age changed the climate, geography, and vegetation available to hunter-gatherers. The last glacial cycle of the Ice Age began about 116,000 years ago. Ice sheets expanded, and sea levels fell drastically. The cold increased and reached its maximum 24,000–16,000 years ago, a period referred to as the Last Glacial Maximum. Average temperatures fell by as much as 57° Fahrenheit near the great ice sheets in the north. The severe cold meant that no trees or forests could grow in Europe and North America; instead, cold dry tundra covered the earth. The landscape of Spain 16,000 years ago resembled that of Siberia today. Soon afterward, the earth began to warm up, until by 13,000 years ago, the Ice Age suddenly ended, and climatic conditions similar to those of today prevailed.

CLIMATIC VARIABILITY AND THE IMPOSSIBILITY OF AGRICULTURE DURING THE ICE AGE: This last glacial period did not just mean cold temperatures; it also meant great climatic variability. Temperature and rainfall fluctuated wildly from year to year, rendering the landscape much less stable. One year, abundant rainfall would guarantee large expanses of wild wheat; in other years, cold, dry conditions meant that only a few of these plants survived. Herds of animals varied their migrations as a result of changing vegetation. Pleistocene foragers could not expect to harvest the same wild plants or hunt the same wild animals each year. They had to be flexible, to adapt to the difficult climatic circumstances.

Climatic variability also meant that establishing agriculture was impossible. Wannabe farmers could not rely on consistently good harvests of any one crop. Hunting large animals and gathering a wide range of plants (whatever was abundant) was a much safer strategy. In contrast, the gentler climate of the Holocene, where more predictable rainfall and temperature patterns meant that harvests succeed, on average, 80 percent of the time, encouraged agriculture. Given the greater stability of plant and animal communities, mutual dependence could easily evolve as part of normal foraging patterns. The extinction of many of the large animals that Pleistocene hunters had previously hunted by this period made seed and root crops look pretty good. If every year you harvest the same grove of wild barley, it is natural to encourage its growth—perhaps spreading some of the barley seeds to extend its natural range, or diverting a stream so that the grove is well-watered. Foragers all over the world began gathering seed and root crops (such as wheat and potatoes) that now occurred in large fields every year. They began to care for the plants they found most useful, keeping these crops free of weeds, or casting their seeds over a wide area. Over time, a reliance on these crops developed into full-fledged agriculture.[4]

OTHER FACTORS BEHIND THE ORIGINS OF AGRICULTURE: Of course, many other factors encouraged the adoption of agriculture. Even before the end of the Ice Age, the majority of the big game animals that human hunters had stalked during the Pleistocene were disappearing. Hunter-gatherers could no longer rely on a steady diet of reindeer; they had to diversify the plants and animals they ate in order not to starve. As a result, many foragers began to eat a *broad-spectrum diet*—relying on quick, hard-to-hunt rabbits, as well as easily gathered turtles, for example—or adding seed plants, which are time-consuming to process, to their diet in addition to tasty berries. Population growth was another factor that probably encouraged people to experiment with agriculture. By the end of the Ice Age, the most favored habitats may have been rather overcrowded.[5]

The timing of these developments, and the exact crops involved varied due to regional conditions. In much of the world, however, domesticated crops appeared soon after the end of the Ice Age. Around 10,000 years ago, farmers in western Asia domesticated rye, barley, and wheat. In northern China, the first grains of domesticated millet appear in simple villages along the Middle Yangtze River 8,500 years ago.[6] The first experiments in domestication in the Americas may date to a similar period, although the evidence from the Mexican highlands and Amazonia is much less certain. In other places, however, domestication did not begin until thousands of years later. Andean foragers, for example, may have domesticated potatoes only 5,000 years ago, although they had undoubtedly eaten wild potatoes (and maybe even cultivated them without domesticating them) for much longer. In fact, the process of domestication is still going on. Strawberries, blueberries, and pecans are all recent domesticates.

ANIMAL DOMESTICATION: Not every plant was worth domesticating. Of 200,000 higher plants, only about 100 have yielded valuable domestics, and we commonly eat even fewer plants. Similarly, farmers and agricultural scientists have managed to domesticate only 14 of the 148 species of large mammals (such as cows, sheep, pigs, water buffalo, and camels). Why is this the case? Why weren't Khoisan cowboys in South Africa riding zebras to round up impalas? Why didn't the Navajo herd North American bighorn sheep, and spin and weave their wool into rugs? After all, the three species of zebras are so closely related to horses that they can interbreed; impalas are caprids, just like the goat (perhaps the first domesticate); and bighorn sheep are not that different from their Asian cousins. The answer is that in order to be domesticated, an animal must meet six criteria: (1) their diet must be easily supplied by people, (2) they must grow quickly and give birth often, (3) they must be able to breed in captivity, (4) they must have a follow-the-leader "herd" mentality, (5) they must not panic in enclosures or when faced with predators, and (6) they must have a good disposition. The Khoisan could not domesticate the zebra (nor could South African agronomists) because zebras are nasty-tempered creatures who "have the bad habit of biting a handler and not letting go until the handler is dead, and thereby injure more zoo-keepers each year than tigers."[7] They did not domesticate impalas because they panic in enclosed spaces and die from heart attacks. Similarly, the Navajo could not domesticate the bighorn sheep because, unlike their Eurasian relatives, such sheep do not occur in herds, and thus will not follow a human leader as they would the herd leader. Generally, animals that could be domesticated were domesticated thousands of years ago. People in both China and the Middle East domesticated pigs, while Africans, European, Middle Easterners, and South Asians all domesticated cattle independently. The reindeer is the only big animal that has been domesticated during the last millennium—hardly the most useful farmyard animal today. Otherwise, recent domesticates include the chinchilla (a fur-providing animal) and the laboratory rat, which are nowhere near as valuable as horses, sheep, or even chickens. In the following section, we will explore how people domesticated both plants and animals in three different regions, in order to see how different regional factors affected this global process.

How It *Really* Happened

THE FIRST GARDEN OF EDEN: WESTERN ASIA

The plants and animals that form the basis of agriculture in Europe, the Middle East, Australia and the United States today—wheat, barley, peas, lentils, pigs, goats, sheep, cattle, rye, figs, olives, and grapes—were all domesticated for the first time about 11,000 years ago on the eastern shores of the Mediterranean and along the Euphrates River. More research has been carried out in the Middle East on the origins of farming than anywhere else in the world. As a result, accounts of the development of agriculture often make farming sound predestined, something that had to happen in a certain way. Of course, nothing in history is predetermined; a lot of random elements affected why and how these crops were first planted. The origins of agriculture in western Asia made use of many inventions. These included both specific tools and social adaptations like sedentarism—staying in one place year-round—which developed before people started intensively using the ancestors of today's crops.

LATE PLEISTOCENE FORAGERS IN WESTERN ASIA: Many of the first agricultural tools used in western Asia began life as something else. Immediately following the Last Glacial Maximum, several hunter-gatherers lived on the eastern coast of the Mediterranean, in present-day western Asia—in settlements that ranged from temporary camps to less ephemeral settlements. These foragers, the Kebarans, had a broad-spectrum diet, which included a wide variety of plants, small and large mammals, birds, and sea creatures. They also used red ochre as a cosmetic, which they made

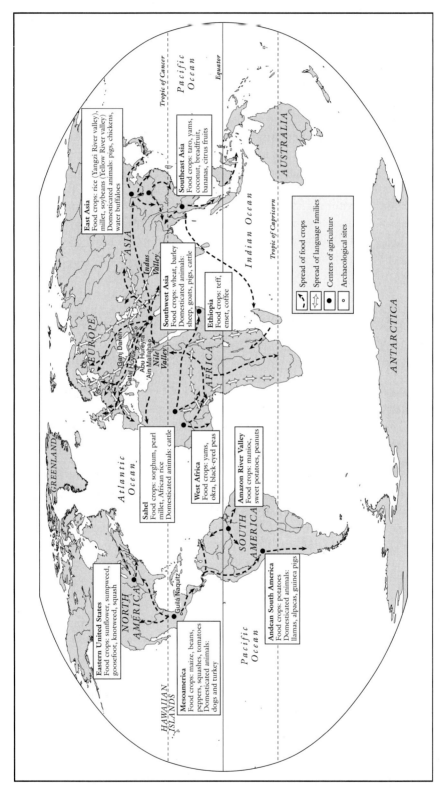

Origins and early spread of agriculture.

into a paste with a grinding stone. Later foragers adapted this old technology and put it to new use, grinding wild wheat and barley grains to make flour and groats for porridge. Similarly, people living in a semi-sedentary village in Mureybet, Syria, made adzes and hoes—not for planting seeds, but as woodworking tools. It was only later that these technological innovations, which so many archaeologists once believed were invented for harvesting wild cereals, were actually used for this purpose.

THE NATUFIAN ADAPTATION: People living in Kebaran settlements probably occasionally harvested wild cereals when they found abundant stands, but they did not use these resources intensively. It was a different story with the people who came after them, the Natufians. The Natufians took harvesting cereal crops very seriously, if the condition of their teeth is anything to go by. Examination of skeletons recovered from 'Ain Mallaha, Israel[8], shows that most of the people buried there had rotten teeth from eating too much barley gruel and wheat flatbread (yeast did not achieve widespread popularity until later). Otherwise, hunters focused on procuring gazelle, while not scorning to collect the birds, fish, tortoises, and shellfish that lived in or near the lake next to the settlement. 'Ain Mallaha is an example of the other important change that immediately preceded the adoption of agriculture, sedentarization. 'Ain Mallaha may have had a year-round population of 200 to 300 people. Although this is a tiny hamlet today, at the time it represented one of the largest human settlements that had ever existed.

A YEAR-ROUND SETTLEMENT AT ABU HUREYRA: Abu Hureyra, located on the banks of the Euphrates River in Syria, was another permanent settlement during this time. Like their contemporaries in the Jordan valley, the people at Abu Hureyra specialized in gazelle hunting and ate a large variety of wild seeds, including wild wheat and wild rye. An intensive study of the plant and animal remains found in the excavation (which took more than 25 years to complete!) proves that people lived there permanently, just like at 'Ain Mallaha.

THE YOUNGER DRYAS AND THE ORIGINS OF RYE AND BARLEY CULTIVATION: Abu Hureyra remained occupied during the Younger Dryas event, a millennium-long return to Ice Age conditions 11,000 years ago (Chapter 1). This climatic downturn caused the environment around the site to become drier and much of the wild food that the Abu Hureyrans had enjoyed for centuries to disappear. Wild lentils were the first to go, followed by wild wheat and rye and other vegetable staples like feather grasses and chenopods. Faced with a food shortage, the Abu Hureyrans decided to cultivate rye. An analysis of the plant remains shows that just as the drought killed off the wild wheat around the site, the inhabitants began to plant domestic rye, beginning about 8,000 BCE. The Younger Dryas event sparked plant domestication all over western Asia. At Netiv Hagdud, Israel, the inhabitants began planting and harvesting barley as a response to the same environmental pressures that led to the development of domestic rye at Abu Hureyra. Bread wheat was probably domesticated somewhere between Mureybit and Jericho in the Levant, while einkorn wheat (a variety we seldom eat today) might have come from Anatolia.

SHEEP AND GOAT DOMESTICATION: Barley, wheat, and rye quickly spread from Syria and Israel into Anatolia and northern Iraq. Some of these early settlements were built in the foothills of the two largest mountain ranges of the Middle East, the Taurus and the Zagros. People who lived in this area had long been familiar with the habits of some of its furry denizens, mountain goats and mountain sheep. Soon after the last Ice Age, at 9000 BCE, a group of complex foragers living at Zawi Chemi Shanidar, Iraq, had started keeping goats in corrals, slaughtering the tender young males when they wanted fresh meat for dinner. The spread of domesticated crops into this area allowed these complex hunter-gatherers to live longer in one place. This, along with their previous corralling practices, may have tempted them to experiment with breeding goats for meat. Selective breeding meant that certain traits—such as intelligence and large size—quickly disappeared from the goat population. An examination of animal bones from a village site in northern Iran called Ganj Dareh shows that by 7500 BCE, domestic goats were here to stay. Sheep were probably domesticated at about the same time. By 7000 BCE, these two domesticates had spread across western Asia, as people got tired of gazelle hunting (and as the gazelle population thinned out).

PIG AND CATTLE DOMESTICATION: Pigs and cows were also domesticated in western Asia, but this process took longer and happened later. Several studies suggest that these animals were domesticated between 7250 and 6700 BCE. Aurochs, the wild progenitors of our docile dairy cattle, are extremely large and can be dangerous. Interestingly, their domestication follows several millennia where they were an important symbol, perhaps of male virility. At Çatalhöyük, a 10-hectare site in Turkey that was first occupied at approximately 6000 BCE, houses and religious shrines have been uncovered giving us a glimpse of Neolithic religion. Neolithic painters captured images of bulls with erect penises, while caches of wild bull bones deposited in unusual areas may be the remains of feasts celebrating the power of these massive animals. Most (if not all) of the cattle remains found at Çatalhöyük were from wild aurochs, however, at about the same time domesticated cattle began appearing in western Asia. The symbolism of cattle is so ubiquitous throughout the Middle East of this period that one theory about the origins of agriculture in western Asia proposes that wheat cultivation and sheep herding (and the rest of the agricultural complex) spread as part of a new religious cult—the way that Arabic spread across the same region, piggy-backing on the success of Islam. Unlike in later Europe or Africa, cattle never supplied a large percentage of meat in people's diets in the Middle East, where goats, sheep, and in wetter areas, pig, tended to dominate. It seems possible that cattle were first domesticated not by a group of beef lovers, but as part of a Neolithic religious tradition.[9]

GIFTS FROM THE CAYMAN: MESOAMERICA

Fields of wild wheat and barley and herds of mountain sheep and goats remain a common scene in the mountains of the Levant, while teosinte (wild corn) grows intertwined with squash and wild beans in Mexico. These plants form the trinity of Native American agriculture, and as a result, most of the research into the origins of agriculture during the last 50 years has told "the story of maize." Yet these were not the only plants that early farmers domesticated in Central America; tropical crops such as manioc, sweet potato, arrowroot, and avocado were also important. There were two centers of domestication in Mesoamerica and Central America: one in the arid highlands of Mesoamerica, where the wild ancestors of corn, squash, and beans are plentiful, and another in the tropical lowlands, which spread from Panama to Amazonia, where root crops still flourish today. Unlike in western Asia, animal domestication was never an important part of the Mesoamerican experience.

FORAGERS AND MOBILE FARMERS IN MESOAMERICA: Botanical analysis shows that modern maize is most closely related to a species of teosinte that grows in the Rio Balsas river basin in Michoacán and Guerrero, Mexico. Unfortunately, very little archaeological research has been conducted there, none of which is concerned with the origins of agriculture. Perhaps protofarmers living along this river experimented with teosinte and produced the first domestic maize. This crop probably quickly spread to other areas of Mesoamerica. Caves excavated during the 1950s and 1960s in Oaxaca, Tehuácan, and Tamaulipas in the Mexican highlands preserve some of the first evidence of maize, squash, and bean cultivation. Highly mobile hunter-gatherers—and later farmer-gatherers—visited these caves for short periods over many thousands of years, beginning about 8750 BCE. At one of these caves, Guilá Naquitz, in Oaxaca, hunter-gatherers exploited a wide variety of wild plants and animals, eating everything from cactus fruit to jackrabbit. This lifestyle required careful planning and constant movement to exploit each seasonal resource fully. People spent most of their time living in small groups, far removed from other foragers, so as not to exhaust the food supply. These hunters and gatherers no doubt manipulated the countryside much like other pre-agricultural people, using burning and selective hunting to assure an adequate supply of their favorite foods around each of the camps where they lived. At some point, this manipulation became plant cultivation, and then plant domestication. Unlike in western Asia, where the first farmers lived in permanent or semi-permanent villages before they began systematically planting crops, in Mesoamerica the first farmers shared the highly mobile lifestyle of previous hunter-gatherers. They sowed seeds at one camp, left, and then returned to gather the ripe crops months later. Farming represented only a small supplement to their diet for thousands of years, as people continued to rely on wild plants and animals for most of their calories.

SQUASH, THE FIRST MESOAMERICAN DOMESTICATE: The origins of farming in Mesoamerica began at roughly the same time as in western Asia, but the focus was very different; the first cultivated crops were not staples (root or cereal crops) that might tempt foragers to give up their mobile ways, but container crops, like gourds, or spices, like the chili pepper.[10] The oldest domesticated plants from Mesoamerica are a few 10,000-year-old squash fragments, useful for both their tender flesh (pumpkins and zucchini) and their watertight rind (for containers).[11] The earliest grains of domestic maize, on the other hand, are only 6,250 years old.[12] As a result of this lack of interest in domesticating staples, in Mesoamerica plant domestication did not lead immediately to village life. Instead, bands of hunter-gatherers (focusing only on wild plants and moving frequently) gave way to "incipient agricultural bands" (people who domesticated plants, but used them infrequently), who gave way in turn to "agricultural bands" (people who moved less and planted more), and finally, to agricultural villages. This last development, however, took at least 2,000 years or, more probably, 6,000 years! The inhabitants of western Asia, China, Pakistan, and Greece, all places where people had developed agriculture at about the same time as in Mesoamerica, had already developed writing, legal systems, political theory, and international diplomacy by the time that the Mesoamericans began living in villages.

RAIN-FOREST AGRICULTURE: Agricultural beginnings in the rain forest, however, show a different trajectory. Excavating early agricultural sites in the rain forest is nearly impossible. In forests, people generally build their houses out of wood and use wooden tools, all of which rot long before archaeologists can find them. Few artifacts survive the constant wetting and drying of this environment. A village of 100 people, consisting of 20 wooden huts with wooden furniture, bowls, decorated basketry, wooden digging sticks, wooden and cloth children's toys, clothing, beautifully woven rugs, and a few stone knives may be just a scatter of broken stones by the time archaeologists get to it.

PHYTOLITHS AND LAKE CORES: Because conventional excavation techniques cannot answer questions about early agriculture, archaeologists have developed new techniques that can. Most typical plant remains, like charred seeds, leaves, and roots, are quickly destroyed in the tropics. Phytoliths, the silica skeletons of plants, however, can survive in the soil, on pottery vessels, on grinding stones, and on other stone tools. Studying lake sediments allows geologists and biologists to investigate changes in the regional environment. Each year, new soil washes into the lake, trapping pollen from nearby trees and grasses, which record changes in the environment over time. Lake cores taken from Central and South America preserve information about the entry of people into the tropics, their pattern of plant gathering, the beginnings of horticulture (planting small gardens in contrast to large fields), and the development of slash-and-burn agriculture. Excavating a single site may give us very precise details about what happened in one place, but analyzing lake cores lets us see the effects of human decisions on a regional scale.

GARDEN PLOTS AND THE ORIGIN OF AGRICULTURE IN THE TROPICS: Agriculture may have begun in the American tropics, probably in northern South America or southern Central America (the evidence for central Panama is especially good), soon after the end of the Ice Age. Prior to this, the first Americans in the tropics had intensively hunted large mammals, which went extinct at just this time (see Chapter 1). After these disappeared, at about 10,000 BCE, people needed to find a new way to live. Tropical plants provided an obvious new resource; Central Americans probably began manipulating their forest environment as soon as the mega-fauna disappeared. The first plants cultivated in the American tropics were grown in crowded garden plots—similar to those which can be found next to houses today. Gardeners planted a large variety of plants, including both domesticated and "wild" species. The first crops were probably tubers such as manioc, sweet potato, and arrowroot, although avocadoes, chili peppers (domesticated in both Mesoamerica and South America), squash, and (later) corn were also grown.

ROOT CROPS AND SLASH AND BURN AGRICULTURE: Two thousand years after the first experiments in plant growing, farmers in the tropics began to rely almost entirely on crops such as corn and manioc (a root from which you can make flour). This probably happened as soon as farmers managed to breed a type of maize that yielded enough corn to make planting large fields of it worthwhile. Maize spread south from Mexico as part of a slash-and-burn agricultural system, uniquely adapted to the infertile soil of the tropics. Even today, small, shifting settlements in the tropics generally practice slash-and-burn agriculture. These settlements are made up of only a few families and do not resemble the nucleated villages of western Asia. Slash-and burn agriculture was so successful, and this social system so stable, that people in the American tropics chose not to settle in villages for thousands of years.[13]

FEW DOMESTIC ANIMALS IN MESOAMERICA: There is one final difference between western Asian and Mesoamerican agriculture that must be considered: the almost complete lack of domesticated animals in the latter. In Southwest Asia, the dog, pig, sheep, goat, and cow were all domesticated within a few thousand years of plant domestication (the dog was actually domesticated even earlier, by hunter-gatherers). A bit later, horses and donkeys were introduced from Russia and Egypt, respectively. In contrast, the ancient Mesoamericans acquired just two domesticated species: the dog and the turkey. Dogs fulfilled much of the same role that they did in other societies—hunting, guarding, and so on, but they were also raised for food. Why did the Mesoamericans domesticate so few animals?

The answer can be found in the impetus for agriculture here, the extinction of many large American land mammals. The animals that remained in both the highlands and the tropics could not be domesticated for one reason or another. It's not that the Mesoamericans didn't try; the tendency to keep pets is universal. Thousands of Mesoamerican children no doubt annoyed their parents as they tried to raise baby tapirs and peccaries. Unfortunately, these animals, when they become adults, are generally solitary, mean, and cannot bond with humans the way dogs, goats, and sheep do. The lack of large animals, especially for traction, is one reason that Mesoamericans never used the wheel for transportation (there was no animal to pull a cart, if they had invented it) and were susceptible to the early Spanish, with their formidable horses, when they invaded.

SAHARAN PASTORALISTS/SAHELIAN FARMERS

Agriculture in Africa represents another quagmire, but one which provides a nice transition between the initial domestication of plants and animals (pristine agriculture) and their diffusion to new areas. Archaeologists who study agricultural origins outside of Africa blithely argue that animals were not and could not be domesticated before plants; Africanists know that in Africa, people living in the Sahara and Sahel domesticated plants long after they had already established pastoral economies. The entire story of African agriculture is complex and little understood. Few early farming settlements have been excavated in Africa. We have a few intriguing facts and many possible hypotheses, but the story could change dramatically in the future.

Saharan Pastoralists: The African story is a curious combination of borrowing crops and techniques from western Asia, responding to environmental crises, and developing new crops suited to sub-Saharan Africa. The story begins, oddly enough, in the Sahara. Many people think of the Sahara as a vast, hostile, uninhabited desert. Even today, people such as the Tuareg and the Berbers live as pastoralists (mobile shepherds) in the Sahara, passing between the Sahel, the savanna directly south of the desert, and the Mediterranean coast. Between 10,000 to 5,000 years ago, the Sahara received more rainfall than it does today. As a result, the desert was smaller, and seasonal rainfall created shallow lakes interspersed with grassland within its midst. Herds of antelope and wild oxen thrived in the Sahara, moving slowly between the lakes. The early people of the Sahara managed these herds of wild oxen, culling animals at will, simply by controlling them as they moved to and from different water sources. After a while, this control turned into domestication. By managing herd migration, these proto-pastoralists encouraged inbreeding that reduced the size of the animal, increased milk production, and improved their temperaments. Herds of cattle, with distinctive coloring and horns (produced through careful breeding), graze in Saharan rock paintings dating to about 5,000 BCE that provide evidence for pastoralism by this point (Figure 2). New genetic evidence suggests that many of these cattle were probably domesticated in the Sahara, whereas other species came from western Asia.

FIGURE 2
(*Top*) Two cave paintings produced 5,000 to 6,000 years ago illustrate the different roles
played by men and women in the early days of agriculture. Here women harvest grain.
(*Bottom*) Men herd domesticated cattle in the early days of agriculture. This painting and
the previous one both came from a cave at Tassili n'Ajjer in modern-day Algeria.

DÉCRUE AGRICULTURE: The cattle herders did not live off steak alone;
instead, other rock art shows goatherding, wild-animal hunting, and normal family life, while the grinding stones that litter their camps suggest
they harvested the fields of wild grain that sprung up near lakes in the
desert. Even today, tasty wild grasses grow in the Sahara; travelers in the

nineteenth and twentieth century described people harvesting them and even selling them in markets, where city people used them to make beer and couscous.[14] They may also have sown wheat and barley, imported from west Asia, along the shores of the shallow Saharan lakes. These lakes expanded with the seasonal rains and then slowly shrank during the dry season, allowing farmers to plant seeds in the moist soil that was revealed as the lakes retreated. These seeds did not require any rain to germinate; they grew entirely from the moisture preserved in the soil. Farmers living in the Nile valley used this same strategy, called *décrue* agriculture. With it, they produced the wheat that made the pyramids possible, and turned Egypt into the breadbasket of the Roman Empire.

THE ARIDIFICATION OF THE SAHARA AND THE DOMESTICATION OF SORGHUM: Unfortunately, beginning about 3500 BCE, the Sahara began to dry up, forcing the cattle pastoralists to move south. Prior to this climatic downturn, cattle herders had rarely ventured south of the Sahara because of the swarms of tsetse flies, which plague cattle and humans with deadly sleeping sickness. Luckily, the sudden aridity also caused the flies to shift their habitat south. After these flies had fled, cattle herders moved into both the East African highlands, where tribes like the Masai still herd cattle today, and the Sahel. The world south of the Sahara presented different problems for farmers than the desert or the Mediterranean coast. Wheat and barley, the staple grains that the Saharan farmers used in the desert, cannot grow in the Sahel. These crops need winter rainfall, maturing during the colder months of the year, and are harvested before the heat of the summer. They were domesticated in and are suited to a Mediterranean-style climate, which has this sort of rainfall pattern. South of the Sahara, on the other hand, closer to the equator, the summer monsoons bring the rains, and there is less seasonal variation in temperature. The failure of their previous staple crops led these Sahelian pastoralists to experiment with domesticating new plants from the savanna. As a result, they domesticated two types of millet, sorghum, and African rice—all grains that grow around water holes. The earliest sorghum found so far is 4,000 years old and comes from Adrar Bous, a pastoral site in the southern Sahara. The Saharan pastoralists could use the same farming technique, *décrue* agriculture, to grow these new crops.

AFRICAN CENTERS OF PLANT DOMESTICATION: When they moved into the Sahel, these pastoralists did not confront an empty world, but instead, one where hunter-gatherers lived alongside early agriculturalists who had probably already domesticated yams and other tuber crops near the Niger River. Unlike western Asia, where there was a definite center for early agriculture, in Africa, crops were domesticated across a broad band of the continent. The Saharan pastoralists domesticated sorghum, pearl millet, and finger millet as they moved into northeastern sub-Saharan Africa, while the inhabitants of the Ethiopian highlands domesticated a slew of other plants: teff, enset, and most popularly, coffee.

INDIAN OCEAN TRADE: People often think of sub-Saharan Africa as isolated from Europe and Asia, at least until the early days of colonialism around 1500 CE. In reality, Africans interacted with people to the north, by way of the Nile River and the Mediterranean, and probably more importantly, with people to the east, by way of the Indian Ocean. Both sorghum and millet, originally domesticated in Africa, made their way, presumably in merchant's ships, to south Asia by about 1000 BCE. Similarly, Asian bananas, yams, and taro grew in gardens throughout sub-Saharan Africa by 1400 BCE, while Asian rice flourished along Africa's east coast. All of these crops were originally domesticated in Indonesia; they traveled the other direction along the trade routes that brought African sorghum to India.

Unfortunately, due to the paucity of archaeological activity in Africa, we do not have a precise chronology for when plants were domesticated in the three major centers of African agriculture—the Niger River, the Sahel, or Ethiopia—nor can we date precisely when Asian crops reached Africa. In fact, most of our evidence about early agriculture comes from the plants themselves. The story they provide is tantalizing, one that we can only hope will be fleshed out in the future.

THE SPREAD OF AGRICULTURE AND THE MAKING OF TODAY'S WORLD

The story of how people domesticated animals and plants in specific locales may seem irrelevant to the way we live today. Far from being the agricultural giants of the modern world, Syria, tropical Mexico, and the African Sahel cannot produce enough grain today to feed their own people. Instead they import grain from today's agricultural centers: the American Midwest, Canada, France, Ukraine, and the Argentinean Pampas. Not one of these places developed agriculture on their own. Instead, these centers benefited from the spread of agriculture, adopting crops that had been domesticated elsewhere and tweaking them so that they grew well in their new homes.

The spread of agriculture had consequences far beyond the creation of these modern breadbaskets. The origins and subsequent extension of agriculture affected everything from language distributions, to epidemic and lifestyle diseases (from measles to heart disease), to overpopulation. Certainly, our world atlas was altered by many subsequent factors—the origins of cities and states, capitalism, and decolonization—but none of these developments would have occurred without agriculture. In a sense, they, too, are the consequences of the decision to plant rather than to gather.

LAMBS, LENTILS, AND LANGUAGES

In 1783, an English colonial official named William Jones was sent to India as a judge. Like most young men in those days, Jones's education had taught him to read ancient Greek and Latin. In India, he amused himself by studying a little Sanskrit, the ancient language of Hindu texts. He was quickly astonished by its similarities to Greek and Latin. Vocabulary words were clearly related, as were grammatical forms. Following Jones's discovery, linguists now accept that most of the languages spoken in Europe (everything but Hungarian, Finnish, Estonian, and Basque) share a common ancestor with the languages of Iran (Persian, Kurdish), and some of the languages of India (Hindi, Urdu). Why this should be so, given the distance between say, Bombay and London, has puzzled linguists and historians for 200 years.

INDO-EUROPEAN LANGUAGES AND THE SPREAD OF FARMING IN EUROPE: Recently, however, an alliance between two geneticists and an archaeologist have suggested the answer. They have looked at the genetic profiles of modern Europeans, North Africans, and western Asians, and compared the results to archaeological models of the spread of agriculture from west Asia into Europe and south Asia. They see it as a spread of people, early farmers seeking new fields, rather than solely one of language. These farmers moved quite slowly, only about 1 kilometer a year. This was fast enough, however, to make Europe a continent of farmers in just a few millennia. This rate wasn't constant; instead, farmers spread quickly along the Mediterranean coast (areas where the wild ancestors of the southwest Asian crops thrived—as did the crops themselves) and more slowly in the interior. The conquest of Scandinavia took a long time because the climate and resources there were initially more suited to fishing than to farming. It seems unlikely that the farmers ever even thought of themselves as moving; instead, the farmer's daughter and her husband would move a few kilometers down the road once they were married. A generation later, their children would follow suit, until suddenly they were scattered all over the continent.[15] Yet, Europe wasn't empty when the farmers began to move in. Instead, genetic variance suggests that these early farmers mingled their genes with those of the early hunter-gatherers, a farmer's son taking a hunter-gatherer bride, and vice versa.

FORAGER-FARMER COMPETITION: Some archaeologists argue that this scenario whitewashes history.[16] Its insistence on peaceful trips to beautiful new forests where farmers fell in love with nubile forager women ignores most of the archaeological evidence. Instead, mass graves, arrows embedded in skeletons, and village defensive systems all paint a grislier picture, one in which the first farmers and the last foragers were locked in a deadly struggle. The migration of farmers into Europe may have been a process of violent conquest that involved the massacre of most of the foragers previously living in the forest and river basins of this continent, rather than a slow, peaceful natural process. The few forager genes that still persist in modern Europeans may be there because the farmers raped and enslaved hunter-gatherer women, not as symbols of multicultural romance.

PROTO-INDO-EUROPEAN IN EUROPE: Archaeological evidence suggests that the farmers who became Europeans originally lived in Anatolia (modern Turkey, part of the western Asian agricultural heartland). Colin Renfrew has suggested that these early farmers spoke "Proto-Indo-European," and that they managed to impose this language on much of the continent, where it gave rise to most European languages. The only area of Europe where people still speak an old hunter-gatherer tongue is the Basque country along the French-Spanish border. Genetic analysis has shown that the people here are not closely related to other Europeans, suggesting that they managed to avoid intermarriage with the foreigners and hold onto their land, language, and culture.[17]

PROTO-INDO-EUROPEAN PASTORALISTS IN IRAN AND INDIA: How then did Indo-European languages spread to Iran and India? Archaeological evidence shows that the early inhabitants of both of these regions spoke other languages: Elamite in Iran and probably Dravidian (a language family now confined to Southern India) in India. Indo-European languages may have followed other agricultural innovations like pastoral nomadism and the domestication of the horse into Iran and India, between 4000 and 1500 BCE. Many archaeologists believe that the first horse pastoralists—nomads who herded sheep, goats, and horses over the great plains of Central Asia—spoke a language ancestral to Persian and Hindi.[18] Various Indian myths remember their migration and exalt the horse, the tool that they used to subdue the earlier farming population.

OTHER THEORIES OF THE INDO-EUROPEAN DISPERSAL: Other archaeologists and linguists have suggested that the spread of horse pastoralism into Europe also established Indo-European languages there. They believe that pastoralists living in southern Russia and central Asia are the original proto-Indo-European speakers, and that they only entered Iran, India, and Europe around 2000 BCE.[19]

RESIDUAL ZONES AND SPREAD ZONES: Eurasia is not the only place where the diffusion of seeds and tools led to the diffusion of language. Indeed, Renfrew divides the world up into two language areas: residual zones and spread zones. Residual zones are places where people speak the old hunter-gatherer languages, where the spread of agriculture has not altered language distribution. Spread zones are the opposite: regions where the spread of farmers, pastoralists, or arctic peoples has transformed the linguistic landscape during the past 10,000 years. Often residual zones are places where people speak an astonishing array of completely unrelated languages, like Papua New Guinea, whose 5 million inhabitants speak 715 languages. Residual zones are not necessarily places that did not experience agriculture, as farmers lived in New Guinea. Instead, they were generally areas where geographical, social, or cultural barriers kept farmers and farming techniques from spreading easily.

THE SPREAD OF BANTU LANGUAGES AND THE AFRICAN IRON AGE: One of the best examples of the confrontation between a variety of residual zones and a spread zone occurred a few thousand years ago in sub-Saharan Africa, when Bantu farmers left their West African homeland, near Cameroon, and started colonizing the area that lay to the south and east. Today, these Bantu farmers, who occupy tens of thousands of square miles, from the Congo to South Africa, speak 500 related languages. That might seem like a lot, but these "different languages" are so similar to each other that "they might be more properly called 500 dialects of a single language."[20] The Bantu farmers may have left their ancestral savanna as much as 5,000 years ago. At that time, they mostly grew wet-climate crops like yams. They moved through the Congo forest much like the first Anatolian farmers moved through Europe, only a few kilometers a year—slowly if you're the one doing the moving, but quickly if you look at it from a longer perspective. These forests had, and still have, other hunter-gatherer inhabitants, such as the pygmies. These distinctive groups no longer have a distinctive language, but instead speak the same Bantu dialects of their neighbors. This shows another way that farmer languages become predominant; hunter-gatherers may adopt them, even when they do not adopt the rest of the farming lifestyle. By 1000 BCE, Bantu speakers reached East Africa. Here they came into contact with the sorghum and millet farmers we met earlier; from these neighbors they picked up these plants, acquired a few cattle, and most importantly, by 700 BCE, started working iron. As Jared Diamond puts it, "with the addition of iron tools to their wet-climate crops, the Bantu had finally put together a military-industrial package that was unstoppable in the subequatorial Africa of the time."[21] Today in sub-Saharan Africa, nearly everyone speaks a Bantu language, with the exception of a few hunter-gatherers who speak Khoisan—who, like the European Basque speakers managed to avoid being engulfed by the Bantu speaking agriculturalists—and the descendants of the people living in the two other agricultural centers of Africa: Sudan and Ethiopia. The Basques and the Khoisan both live in naturally isolated areas, places where Indo-European and Bantu speakers either could not or did not want to live, whereas people from other agricultural centers were able to resist adopting a Bantu language.

CHICKEN POX, COW POX, AND RINDERPEST

HUNTER-GATHERER DISEASES: During the long Paleolithic, hunter-gatherers mostly died from accidents (or the infections which set in afterward) or starvation, not infectious diseases. As Roy Porter, a historian of medicine, puts it:

> Our previous ancestors really were free of the pestilences that later ambushed mankind. Their bodies may have been malformed, arthritic and lame—and prey to gangrene, botulism, anthrax and rabies—but lethal epidemics such as smallpox, measles and flu must have been virtually unknown.[22]

Paleolithic foragers did not suffer from colds, their children never got chicken pox, and no flu epidemics ever broke out among small mobile tribes. All these diseases result from farming (and civilization).

HEALTH OF EARLY AGRICULTURALISTS: A comparison of the skeletons of the last hunter-gatherers (the Natufians), the first farmers and later farmers in Israel and Palestine gives us a time frame for this health decline. In general, the Natufians and the early agriculturalists were quite healthy; in fact, people living in settlements immediately after the domestication of animals were taller, lived longer, and were less likely to suffer from malnutrition than the foragers before them.[23] This was especially true for the men, whose average life expectancy increased from 35.5 to 37.6 years.[24] Following this period, however, average health quickly deteriorated. This decline was related to chronic disease, not to food shortages. It seems that initially keeping tame animals near at hand made for happy and healthy humans; this meat, and recently domesticated peas and beans, meant that the first farmers had a healthy, protein-rich diet. So what happened? How and why did these robust early farmers whose worst health complaints were rotten teeth and bad breath become weak, short-lived villagers in just a few short millennia?

ANIMAL DOMESTICATION AND THE ORIGINS OF INFECTIOUS DISEASES: Those submissive sheep, pigs, goats, cows, and chickens, which provided early farmers with a reliable supply of meat and milk, also provided new vectors for diseases. It seems likely that there was a time lag between the domestication of animals by humans and the domestication of humans by microbes; during this time lag, those first humans probably flourished. Afterward, farmers had to contend with different diseases, depending on which animals they lived with regularly, smallpox, if they kept cattle, and chicken pox and flu if they enjoyed roast chicken.

CROWDED CITIES AND EPIDEMIC DISEASES: Yet diseases from domestic animals did not really begin to wreak havoc on the world until farming villages grew into cities. Suddenly, thousands of people lived in tiny spaces, sharing small rooms with each other, and often the family cow or sheep. In such conditions, epidemic diseases, derived from animals, could attack masses of humans. Many epidemic diseases, like measles and influenza, require a certain number of human victims to remain as a threat in the population. A few thousand is too few. Measles, for example, requires about 500,000 people. By the time the first civilizations got around to inventing writing and large crowded cities, conditions were prime for measles epidemics.

MEASLES, SMALL POX, AND MORTALITY: Today, we think of measles as a not particularly severe childhood disease, yet it can be disastrous for populations that have not developed immunities to it. In 1875, a Fijian chief brought measles to Fiji as a souvenir of his Australian trip. This disease killed 25 percent of all Fijians then alive. Similarly, in the 1960s, increased mining, Amazon exploration, and anthropological work among the Yanomamo in Brazil and Ecuador introduced measles to this group. Up to 40 percent of the population died, despite the fact that anthropologists and public health officials were desperately trying to immunize the population at the same time. Smallpox probably emerged simultaneously. Until the nineteenth century in Europe, when vaccination began, many people bore disfiguring scars from surviving smallpox, after millennia of experience with the disease allowed them to build up immunities. When smallpox reached the Americas, the consequences were devastating. In 1520, smallpox came to the Aztec empire via one infected man arriving from Spanish Cuba. In the next year, perhaps 50 percent of all Aztecs died of the disease, including the emperor. It is doubtful that the Spanish conquistadors would have been so successful, so quickly, without the help of European germs, coming from a long familiarity with domestic animals and large cities, a familiarity that the Aztecs, because of the small numbers of domesticated animals in Mesoamerica, did not have.

Agriculture also contributed to the spread of deadly diseases by unintentionally "domesticating" other animals, such as malaria-carrying mosquitoes and disease-ridden rats. As sub-Saharan African farmers cleared forests and savannas to provide new fields for their millet and sorghum, they unintentionally created ideal environments for anopholine mosquitoes. Several forms of malaria developed only after agriculture brought mosquitoes into intimate contact with people. Similarly, farming helped bring yellow-fever-carrying mosquitoes within the reach of humans. Both diseases were once prevalent over much of the world, including Europe and North America. Now they threaten much of Africa, Asia, and South America. Irrigation agriculture encouraged other diseases, such as schistosomiasis, which thrives in manmade ponds and irrigation canals, or river blindness, carried by flies that breed in rivers or canals.

VILLAGES, WASTE, RATS, AND DISEASE: Urban lifestyles also intensified diseases that first appeared with villages (usually of agricultural origin). Most diarrheal diseases come from accidental contamination of water or food supplies. Moving frequently allows humans to get away from their waste before it can make them sick. Settling down in villages, on the other hand, means that villagers can't leave each time they and their animals foul a site. Instead, infected water and food sickens people, especially infants who do not have the immunities that come from drinking such water for long periods of time. Additionally, villages and, later, cities serve as a haven for animals, such as rats, that thrive upon human garbage and stored grain. The fleas and lice that live in rats' fur have produced many of the worldwide plagues, including the medieval Black Death and typhus. So many risk factors meant that death rates in cities were higher than birth rates. Historians have called cities the "graveyards of mankind," precisely because urban growth relied on a constant influx of people from the countryside.[25]

TOWARD THE 6 BILLION MARK

Agriculture may have unleashed a multitude of deadly diseases, but it has also produced a world population that keeps increasing. Paleolithic foragers may have been fit and well nourished, but their population growth rate was only .02 percent. By contrast, population estimates during the Neolithic are much higher, suggesting that population grew at .1 percent a year. How can we explain this paradox of increased population growth coupled with increased illness? Why does population continue to increase, despite wars that kill millions, frightening new diseases, and widespread famine in various parts of the world? How has the world's population tripled from 2 billion in 1950 to 6 billion just 50 years later?

Anthropologists do not agree on the factors behind population growth. Some argue passionately for farming moms who preferred bigger families than their foraging forebears; others argue that agriculture prevented the death checks that contributed to such low population growth in Paleolithic populations. Both may be right, and population growth could have been a matter of slightly more fertile parents, and slightly fewer famine deaths.

FORAGING AND FERTILITY: Studies of birth rates of contemporary farmers and hunter-gatherers usually show that farmers have more children. Hunting-gathering women often have trouble finding year-round, easily digestible food to wean their children. As a result, they breastfeed their babies for much longer than farming women—often three or four years. Breastfeeding, especially when combined with high levels of physical activity and a low-fat diet, causes women to stop ovulating and become infertile. If a forager woman still manages to get pregnant while nursing, she may practice abortion or infanticide. Many cultures do not make the distinction between abortion and infanticide that Western cultures do. Rather than seeing birth as the dividing line between illegal infanticide and legal abortion, they may set the dividing line at an infant's first birthday, or some such date. Forager women may abort because it is nearly impossible to carry two young children every day while gathering food, and even more difficult to walk 15 kilometers with them every two weeks when her tribe changes camp. The birth of the younger child endangers the health of the mother and of the older child. Biological and cultural factors combined so that Paleolithic foragers had fewer children, more widely spaced, than their farming descendants.

WEANING, EARLY MARRIAGE, AND INCREASED FERTILITY AMONG FARMERS:
Farming women are more fertile today due both to shorter periods of
breast-feeding—grain and animal milk are perfect weaning foods—and
sedentary living. Farmers do not have to travel long distances and can
more easily combine their activities with caring for multiple small children.
Additionally, farmers often marry younger (and may begin menstruating
sooner) than hunter-gatherers. This might be due to the stress of mobility
on young pregnant women, who miscarry more easily than farming girls
of the same age who don't spend as much time on the road. All of these
factors mean that a comparison of simple farming populations and
hunting-gathering populations based on worldwide data shows that farming women have 6.3 children, whereas foraging women have 5.3 children
on average.[26] Such changes in fertility, even coupled with much higher
death rates due to disease, would still fuel the population increase. Some
evidence from the transition to agriculture supports this scenario. Although
average life expectancy overall rose with the origins of agriculture, probably because of better nutrition, it declined precipitously for women, from
35.5 to 30.1 years, probably due to more women dying in childbirth.[27]

HIGH HUNTER-GATHERER MORTALITY: Not everyone agrees with this scenario. Some anthropologists argue that modern hunter-gatherers are a bad
proxy for ancient populations; they live in more degraded environments
and are probably more mobile than most foragers at the end of the Ice
Age. In fact, their manipulation of population data suggests that farmers
and horticulturalists (simple agriculturalists) often have the same number
of children on average.[28] They argue that farmer child-power is a myth,
perpetrated by the same anti-farming, pro-foraging anthropologists that
saw San life as idyllic rather than nightmarish (Chapter 1). They also argue
that we may be overstating the case by dwelling so much on disease and
agriculture-induced cavities. Hunter-gatherers also face disease (though
not epidemics) and are very accident prone. Agriculture increased population by decreasing the risk of famine and allowing more people to live on
the same amount of land. In most places, one hunter-gatherer needs between 1 and 10 square kilometers of land to feed himself, whereas a farmer
only needs one-third to one-three hundredth of a square kilometer. This
meant that when farming emerged, the world could support many more
people than it had for millennia. Suddenly, to the new farmers, the entire
world looked like the empty Americas did to the first settlers across the
Bering Strait. In the same way that the first Americans quickly expanded to
fill the continent, farmers expanded to fill their "empty" world.

But What About the Rest of the World?

Scholars who study complex states are used to seeing agriculture as the necessary building block to institutions like chiefdoms or states. Yet several peoples did not develop agriculture on their own, or actively rejected it for thousands of years. These societies included not only small hunter-gatherer bands living in marginal areas (like deserts and the Arctic) where agriculture was impossible, but also settled hunter-gatherer societies that subsisted from intensive fishing and evolved complex political structures in areas uniquely suited to agriculture. We will consider three examples of nonagricultural societies that resisted agriculture, often until recently, in Australia, California, and Scandinavia.

WHY AGRICULTURE NEVER BEGAN IN AUSTRALIA

By 1 CE, 9,000 years after the first agricultural experiments, every continent in the world had developed or adapted to some form of agriculture. Eurasia had three centers of plant and animal domestication, as did Africa, while North America, Mesoamerica, and South America had one or more as well. Even the inhabitants of New Guinea had either developed agriculture or adapted southeast Asian plants and animals (pigs and chickens) to the high altitude growing conditions of the island. There was only one exception to this general pattern, the continent of Australia. By 1 CE, foragers in central and Eastern Australia had just begun gathering wild seeds and grinding them to make gruel, a cultural adaptation that the Natufians began 12,000 years earlier. Some archaeologists argue that native Australians like these seed gatherers were moving toward plant domestication, an experiment that was interrupted by the British conquest. Nevertheless, when the first British explorers arrived on Australia, it was a continent completely populated by hunter-gatherers. Why didn't native Australians develop agriculture on their own or adopt it from their neighbors in southeast Asia or New Guinea?

ENSO and the Lack of Agriculture in Australia: Scholars have focused on geographic factors to explain why agriculture did not emerge in Australia. Some have argued that Australia lacked good plants and animals to domesticate. Yet taro, arrowroot, and yams, all of which were domesticated in southeast Asia or New Guinea, grow in northern Australia (where foragers gathered them), while in eastern and northern Australia the Aborigines harvested wild millet, a plant related to the staple that the Chinese domesticated. Australia's nonadoption of agriculture might also be explained by its climatic and geographic situation, which meant that it never really experienced the benefits of decreased climatic variability during the Holocene that other continents did. Australia's climatic system is governed by ENSO (El Niño Southern Oscillation), irregular multiyear cycles that involve changing seawater temperatures and rainfall. This means that this continent did not enjoy the usual annual round of seasons that the other continents did. The irregularity induced by ENSO made the Australian climate more similar to Pleistocene climates in terms of dryness and variability than that of other continents. This might explain why Australians did not invent agriculture. Other areas whose climate is also affected by ENSO, such as the Peruvian coast and California, also did not develop agriculture on their own. Like Australia, California never even adopted agriculture before European colonization. The Peruvian coast did, but only long after their neighbors in the mid-Andes, high Andes, and Amazonia had.

WHY DIDN'T AUSTRALIANS BORROW AGRICULTURE?   But this scenario does not explain why Australia never borrowed agriculture from its neighbors. Archaeologists have argued that despite some indirect communication with their New Guinean and Indonesian neighbors, Australians never came in contact with their crops. Indonesians regularly visited Australia beginning about 1500 BCE to collect sea cucumbers, a Chinese delicacy, but the areas where they landed were dry and unsuited to Indonesian agriculture. Similarly, although Australians and New Guineans occasionally traded and met (often on one of the string of islands between these two land masses), Australians never met New Guinean farmers who lived in the highlands. Instead, the island dwellers were generally hunter-gatherers, who may have occasionally raised and eaten pigs and dogs. As a result, Australians, living in areas where Indonesian and New Guinean agriculture would have flourished, never came into contact with these practices, despite trade with both areas.

AGRICULTURE IN PAPUA NEW GUINEA: Yet until about 6,000 BCE, Australia and New Guinea formed one continent, called Sahul. By the time rising sea levels separated the island from the continent, highland New Guineans had already become farmers. Different groups of native Australians came into contact with farmers on myriad occasions during the past 9,000 years. They certainly knew that farming existed, and certain plants and animals were even exported to Australia, where they became feral. Trees like *Canarium*, sago and *Artocarpus* were domesticated on the north coast and islands of Papua New Guinea and then introduced into northern

Australia. Similarly, "wild" Australian yams may have originated in Papua New Guinea.[29] Native Australians had no qualms about using these plants; they were happy to gather their fruit, but they did not cultivate them. The New Guineans even sent over some of their dogs, which became the dingo, a semi-wild animal. Native Australians occasionally befriended these dogs, sleeping next to them on cold nights to preserve body warmth, but they neither ate them nor used them to hunt as did people elsewhere in the world.

AFFLUENT AUSTRALIAN FORAGERS: The reason for Australia's rejection of agriculture may have been partly ideological. Today, after being exposed to transplanted European farming economies, some native Australians continue to reject an agricultural lifestyle. The Gidjingali still gather traditional food despite the fact that they receive social security payments or hold jobs like other Australians. They enjoy foraging and prefer the taste of wild food to domestic food.[30] Contrary to our usual stereotypes about foragers, native Australians may have not adopted agriculture because they lived in a place with abundant resources. They may not have seen how the occasional cultivation of yams in Queensland, particularly given the rate of failed harvests due to ENSO, could actually improve their lives.

CALIFORNIA: AFFLUENT FISHERS AND ACORN EATERS

In the eighteenth century, the British, Russian, and Spanish raced to explore and lay claim to the Pacific coast of North America. The British quickly established the Hudson Bay Trading Company in today's British Columbia, while the Russians had enclaves in Alaska and Bodega Bay, California. The Spanish laid claim to Baja and Alta California, but sailors from all three countries raced up and down the California coast, trading with the inhabitants for fur. When the Europeans arrived in California, the area had a population of at least 400,000 hunters—which may seem small compared to today's population of 34.5 million—but represented one of the densest populations of foragers in the world. Like people in Australia, the inhabitants of California had never bothered to adopt agricultural techniques from their neighbors in the American Southwest, who were adept at growing the imported Mesoamerican trinity of corn, squash, and beans. This might seem strange because today, due to dams and irrigation, California represents one of the most important agricultural areas in the world. If California seceded from the U.S. tomorrow, it would be the seventh richest country in the world. A good deal of that wealth comes directly from the state's agricultural sector.

MANAGING THE NATURAL ENVIRONMENT IN CALIFORNIA: We might see California as a real garden of Eden, one whose agricultural potential is more promising than the actual one in the Mesopotamian desert. The original inhabitants also viewed their environment as plentiful, so plentiful that they never saw the need to toil in the fields when there was so much for the taking. Rather than importing farming techniques from elsewhere, or domesticating the available plants (assuming that there were any native plants in California that could be domesticated), Californians managed the natural environment. Throughout the state, native Californians planted and tended wild oak trees. In the fall, they gathered and stored these acorns, using water to leach out the poison before grinding them into flour. One archaeologist has calculated that wild acorn harvests in California could feed 65 times more people than were actually living in California at the time of its greatest population boom.[31] They also ate the abundant seafood provided by the Pacific: shellfish, small fish (like anchovies), and larger ocean fish, like tuna. During breeding season, the state's rivers were choked with salmon and trout. As in Australia, El Niño events caused the wealth of marine resources to fluctuate according to changes in sea temperatures. These same temperature variations affected acorn and wild grass yields.

CHUMASH CHIEFDOMS: Even without agriculture, the Chumash Indians, who lived in what is now Santa Barbara, formed large villages of up to 1,000 people that fed themselves simply by fishing and practicing "advanced" hunting-gathering techniques. Chumash society was not egalitarian, but possessed an inherited nobility. They poured resources into holding elaborate ceremonies and ensuring that they had the best military in California. Constant raids on other sedentary villages kept them in practice. In short, these hunter-gatherers were as "complex" as many agricultural villagers, and more complex than some of the simplest, including the first farming societies in western Asia. Evidence like this shows that we can answer the question of why people in California did not adopt agriculture by simply pointing to the affluence of these foragers' lives. Why bother to waste your time irrigating corn when everything you want is freely available?

THE SCANDINAVIAN SEA: AN ALTERNATIVE TO AGRICULTURE

We have already explored the quick transition to farming in much of temperate Europe in our discussion of language spread; however, agriculture was not adopted across Europe. Farming spread quickly across southern and central Europe, but northern Europe, especially Scandinavia and the Baltic, resisted the transformation for thousands of years. This means that although early farmers arrived in places like Moldavia and Belgium only about a century apart (despite the 3,000 miles involved), it took them 1,500 years to travel only a few hundred miles north into northern Germany and Scandinavia. In fact, parts of Finland remained the province of hunter-gatherers until the sixteenth century CE; some scholars argue that Finland has yet to fully complete the transition from hunting to farming even now. Why did these early Scandinavians only grudgingly accept the Near Eastern imports of wheat, barley, oats, sheep, goat, cattle, and pig long after the rest of the continent succumbed? The cold climate and short growing season of Scandinavia meant that agriculture required a lot of work for little return. In contrast, the rich sea resources of this peninsula meant that the inhabitants could live off the sea.

COMPLEX HUNTER-GATHERERS IN SCANDINAVIA: Hunter-gatherers first settled Scandinavia and the Baltic states in the early Holocene (at approximately 11,000 BCE), after the glaciers had retreated and forests sprang up in Northern Europe. Originally, they kept to the shores, leaving settlement debris in the form of huge piles of shells, which archaeologists call middens. In Denmark and southern Sweden, people lived in fairly permanent settlements, by the sea. They moved inland, occasionally, to hunt large animals, but generally subsisted on fish, shellfish, and marine mammals.

Like the Australians, these Scandinavians traded with early farmers, acquiring cooking vessels and stone axes, which probably served as prestigious decorations in this cultural context. They might even have traded some furs for a few bushels of wheat—and enjoyed the exotic gruel that they could make out of such an outlandish product—but they did not waste time planting it. After all, depending on the time of year, oysters, tasty seals, shoals of fish, berries, nuts, and venison were available—why bother with agriculture?

SEA CYCLES AND THE ADOPTION OF AGRICULTURE: At least, this seems to have been the Danish mindset until around 3200–3100 BCE, at which time, after having scorned farming for 2,000 years, the Danes suddenly became farmers. Archaeologists offer two explanations for this transformation. The first theory proposes that the sea, responding to a poorly understood climatic cycle, suddenly became less salty. Unfortunately, oysters, a critical food for Danish foragers, require high degrees of salinity. They disappeared, as did other seafood, leaving the Danish foragers increasingly hungry. It seems that the Danes and Swedes reacted to the threatened famine by experimenting with farming. This was initially just a stopgap "to compensate for the decline in marine production."[32] The first farmers tried to integrate agriculture into their previous foraging patterns; unfortunately, farming spoiled hunting-gathering time allocation. To live well, hunter-gatherers must carefully schedule their activities so that they gather and dry berries in the summer, collect nuts in the fall, and go seal hunting in the winter. Fall was already a busy time of year for the Danish foragers, as it was the only season to collect nuts and the best time of year to hunt elk, deer, and other large animals. Unfortunately, autumn, harvest time, is also the busiest time of year for farmers. The more time taken up by wheat harvesting the less time there was for wild nut collection. This caused these early farmers to rely more and more on farming, as they were less able to combine it with their traditional foraging activities. Hunting continued to be important after the adoption of farming in Denmark, but skeleton analysis shows an abrupt change from a marine-dominated diet to a grain-dominated diet. These complex, affluent hunter-gatherers were no more.

THE ABANDONMENT OF FARMING IN SWEDEN: Or were they? At least in middle Sweden, the foragers got another chance after about 2700 BCE. There, deteriorating climatic conditions meant that farming villages suddenly disappeared and were replaced by villages that relied heavily upon fishing and seal hunting. Unfortunately, these foragers lasted only for about 400 years, by which time farming immigrants from southern Sweden succeeded in forcing them onto the outer Swedish islands, where not enough species were available to allow their complex hunting-gathering strategies to survive.

FEASTING AND THE ADOPTION OF AGRICULTURE: Not everyone sees the disappearance of Danish oyster bars as the factor that sparked the adoption of agriculture in Scandinavia. Other archaeologists are more cautious, arguing that the long period of contact, trade, and experimentation with farming (1,500 years, and 500 years respectively) suggests that social factors might have been behind this transition. Danish foragers were relatively affluent; this allowed them to acquire and show off their wealth. The presence of agriculturalists to the south, with their exciting new trade goods, provided foragers with a way of one-upping their neighbors. Suddenly, individuals started hankering after large, private tombs filled with objects that showed they were important to their community. Charismatic leaders began giving elaborate feasts featuring not just boring oyster and seal, but also imported beer (made with grain acquired from the south) and perhaps poppy seed cakes. The Danes adopted farming not for environmental reasons, but because it became the only way to keep up with the Joneses.

THE ORIGINS OF FAMILY, PRIVATE PROPERTY, AND THE VILLAGE

The new opportunities available at the end of the Ice Age encouraged people to abandon a mobile lifestyle for a sedentary one. Sedentary life often accompanies, but is not necessarily dependent upon, agriculture. We have already seen how farming is separate from village life. Danish hunter-gatherers lived in sedentary villages (as did foragers in the Pacific Northwest), while early farmers in Mesoamerica continued to move their temporary camps frequently. In some ways, the transition to permanent settlement was a revolution with consequences as important as the emergence of agriculture. Living in the same place year round led to changes in political, economic and family relationships. First, it required the invention of new community concepts in order to diffuse the tensions that arose because individuals could no longer move away if they were angry with others. Second, it led people to invent new institutions—food storage and property relationships—which allowed them to manage scarce resources and escape starvation. Third, settling down changed gender roles and the balance of power within the community. As a result, sedentarism, coupled with agriculture, led to the invention of the nuclear family and the village, a social organization that laid the groundwork for more complex societies.

INVENTING THE VILLAGE

Settling down, living in the same place permanently, is not the same thing as inventing the village; however, it is often the first step in establishing one. Before the Holocene, all people were basically mobile. In western Asia, the wild-seed using Natufians were the first to live in year-round settlements. Unexpectedly, the development of domesticated crops did not transform the sorts of settlements people built; the round houses of the early Neolithic look like those of the Natufian. In turn, they resemble the sort of temporary shelters that mobile hunter-gatherers construct around the world. As late as 1959, the Hadza in Tanzania built similar huts out of brush in their temporary camps. This suggests that there was a time lag between the development of sedentarism, and of agriculture, and the final social consequences of these revolutions. The earliest social systems of these sedentary communities closely resembled those of mobile hunter-gatherers. In western Asia, it was only 2,000 years after the Natufians first settled down, when farmers had incorporated both domesticated plants and animals into their economy, that people invented the village—a new type of society made up of nuclear families, which was very different from earlier social arrangements.

HOUSING COMPOUNDS AND EXTENDED FAMILIES: Settlements like Netiv Hagdud, Ain Mallaha, and Abu Hureyra looked much the same. Small circular huts, just big enough to house one person (or perhaps a mother and an infant), were arranged in a loose circle. Grain silos or hearths were located in the spaces between each hut, while the empty space in the middle was where people got together to work and socialize, to grind grain, mend tools, or gossip.[33] All of the members of a compound worked together, taking turns hoeing the fields or hunting gazelle. Most property was probably held in common; no one person had a monopoly over grain or the land needed to grow it. Just like in many mobile hunter-gatherer groups, the average population of such a settlement might have been around twenty people. There was probably no concept of a "nuclear family" in these early farming compounds, which were not villages, but merely collections of houses belonging to extended families. Such an extended family might contain two or three brothers, with their multiple wives and children (including, perhaps, married sons).[34]

These housing compounds defined communities in western Asia, from Syria to the Negev desert in Israel for at least 2,000 years. Beginning at about 8500 BCE, the sorts of communities people built changed. Many historians see this period, about 1,000 years after the first plants were domesticated, as representing the real "Neolithic revolution." It was at this stage that people stopped just hunting gazelle and began keeping herds of sheep and goats. Settlements and individual buildings increased in size. House shape also changed—the individual huts of the first farmers gave way to larger rectangular houses, which probably housed nuclear families. They also developed new stone tools that emphasized weaponry and began worshipping their ancestors and agricultural gods. Each of these transformations probably resulted from a complete transformation in western Asian society, with a new emphasis on extended families, property, and male virility.

ABU HUREYRA AND RECTANGULAR HOUSES: We can understand how some of these changes occurred by looking at how life was organized at Abu Hureyra, Syria, during this period. Following a 500-year gap, this site was reoccupied. This new settlement looked completely different from the earlier one. Instead of a few ephemeral pits, the new dwellings were rectangular mud-brick houses complete with black plaster floors, occasionally painted with red designs. The new Abu Hureyra—and its sister communities across the Near East—were not just compounds housing an old hunter-gatherer workforce. Instead, they consisted of many nuclear families, each of which lived in its own house, with its own storage units. The rectangular shape of their houses meant that it was easy to add on (or block off) extra rooms as families grew or shrank. The first agricultural settlements had each contained a single work unit, even if they were made up of what we would consider many different families. With the birth of the village, this changed. Now each village contained several distinct work units/families. Given their lack of cohesion, these new communities had to devise a strategy to unite the different families. Each village had to invent a village identity which all its inhabitants could share.

Beidha: Privacy and Community: A village called Beidha, in Jordan, shows two trends in early agricultural villages. First, people began building their houses with privacy in mind; they made sure that the entrances were hidden and that public space (village lanes, or common squares) was clearly delineated from private space (houses and courtyards). At the same time, buildings with special functions, like community meetinghouses and religious centers (for performing rituals associated with the dead), increased in importance.[35] The community meetinghouse may have been a place where everyone in the community could come together to work out problems, a place where disputes could be solved before they became murderous.

Why did people abandon compounds in order to build villages? How did agriculture and sedentarism lead to the village, and why was there such a long time lag? As we have already seen, both sedentarism and agriculture lead to population growth. This was particularly true in early agricultural societies, before the scourge of epidemic diseases took hold. As settlements grew, and as more and more people set up their little round huts on the outskirts of the mother settlement, the entire social fabric holding together the housing compounds disintegrated. Suddenly, "not every family considered itself closely enough related to its neighbors to be willing to share the risks and rewards of production."[36] Like the old tale of the frugal country mouse and the spendthrift city mouse, families began to privatize their storage so that they didn't have to share the hard-earned fruits of their labor with their lazy second cousins: "and for the rest of the Neolithic period, families who were willing to work harder and store their products privately began to outdistance their neighbors economically."[37]

The transition from small settlements, characterized by extended families and their one- or two-person huts to tidy agricultural villages consisting of nuclear families, has occurred across the world: in Mesoamerica, Western Asia, the Caucasus, the U.S. Southwest, and Egypt. Of course, it has not necessarily followed the same path everywhere. In Nigeria, Tiv horticulturalists continued to live in housing compounds made up of extended families (usually men with several wives) during the twentieth century. In pre-Dynastic Egypt, villagers lived in large circular houses that mimicked the rectangular houses found at Abu Hureyra in function (they belonged to nuclear families), if not in form.

LAND, ANCESTORS, AND PROPERTY

If the origins of the village lie in an aversion to sharing with shirkers, that means that one of the key changes brought about by sedentarism was as economic as it was social. It forced people to develop new ideas about belonging—both about what they belonged to as well as what belonged to them. In most mobile groups, ownership is fairly limited. You might "own" a favorite bow and arrow or a small bead necklace, but no one owns the forest where the wild nut groves thrive or the herds of wildebeest that frequent the steppe. Although groups have complex ideas about where they are allowed to hunt or gather, such concepts are much more flexible than formal ownership. When people began to settle, however, whether or not they become agriculturalists, this changed. Villagers needed to own the means of their own support. They would not tolerate strangers, who might steal their fields or take the salmon from their streams. Such resources could not belong to the earth as a whole any longer; they had to belong to an individual group. The same thing happened with food. The development of the village meant that food no longer belonged to the settlement as a whole; instead, it belonged to each family, who kept their grain in private pantries rather than a communal granary. In one sense, this was simply an extension of foraging behavior, where meat from the hunt must be shared out to nonhunters, but where a gatherer may eat whatever she finds, without sharing.[38]

HOUSES, LAND, AND ANCESTORS: Once invented, ownership proved a rather flexible concept, one which villagers quickly expanded. The first proud homeowners may have been the sedentary Natufians. Along with ownership, concepts of inheritance surfaced. Some Natufians were buried beneath their round huts, perhaps so their children could justify their ownership of the building—the bones of their ancestors acting as a counterpart to a modern deed. If a house belongs to you, and the grain from a field belongs to you, then it is easy to posit that the field belongs to you as well. Across the world, garden plots have been distributed to people depending on the size

FIGURE 3
A plastered and painted skull from Jericho. The plaster remodeling of the features of
human heads is found at several early Neolithic sites in southwest Asia and may reflect an
increasing reverence for ancestors.

of their houses. In England, people can claim "allotments" to this day—
garden patches on the outskirts of town, based upon house ownership.

THE WESTERN ASIAN SKULL CULT: Burial practices at the site of Jericho,
Israel, show the relationship between the creation of ancestors and of
property. Beginning during the Natufian period, the skulls of some adults
were removed long after death, covered with plaster and given new "eyes"
made of shells; the result is a haunting portrait of the dead (Figure 3).
Such skulls were buried in caches, in the southwest corner of a house or
cave, possibly for ritual purposes. This treatment, which was probably
seen as a way of honoring the dead, was given only to certain members of
society, perhaps the most important members of different families. Ancestor traditions—honoring or even worshipping certain ancestors—often
help cement claims of territoriality. If your ancestors have lived in a place
from time immemorial, and you can prove it by pointing to a line of skulls
in the village meetinghouse, then the place where they lived belongs to
you and your children. The skull cult also helped create a village identity
that would defuse tensions by extending the notion of the ancestor, until
everyone in the village, even those who were not closely related, felt that
they "belonged."

Similar skulls have been found at towns from northern Iraq to Cyprus, which shared a common religious tradition. Archaeologists have emphasized that a powerful ideology behind these newly agriculture-reliant societies may have allowed them to disseminate this ideology and the cultural practices that went with it (agriculture). In a sense, these communities exported their culture and ideology: sheep, goat, wheat, and barley probably signaled virility and power to the people outside of the original Fertile Crescent, much the same way that Levi jeans meant prosperity and freedom to Eastern Europe in the 1980s.

BECOMING STAY-AT-HOME MOMS

EARLY VILLAGE MATRIARCHIES?   What did agriculture, the village, and the invention of the family mean for women? Few people pondered this question until the eighteenth century. They believed that society had always been patriarchal (male-dominated) because that was the way it was described in Genesis. Yet the discovery that some Native American societies traced their descent through the female line and held women in high regard challenged this assumption. As a result, various nineteenth century jurists, anthropologists, and political theorists, including Friedrich Engels, believed that a matriarchy, when women held political and social power, had preceded the establishment of the patriarchy and the state. Many feminist scholars in the twentieth century believed that all early agricultural societies in Europe and Asia were peaceful matriarchies that had been overrun and destroyed by warlike, horse-riding invaders.

STUDYING GENDER ROLES: Nowadays, new archeological evidence and different intrpretations have meant that this reconstruction is no longer accepted without question. In almost every case, village life did change the position of men and of women in society, but these changes did not always occur in the same manner. In some places, becoming sedentary led to decreased status, trapping women in unimportant roles. In others, settling down made life easier for women; their new role as guardian of the hearth may have given them symbolic power. Unfortunately, it is difficult to locate individual men and women in the archaeological record. Stone tools, ceramic vessels, and houses do not contain labels telling us the gender of their manufacturer. Instead, scholars make hypotheses according to ethnographic analogy, assuming that, for instance, because men usually make stone tools in twentieth century hunter-gatherer or horticultural groups, they probably made them in the past as well. Yet for every ethnographic rule there is an exception, and the situation during the past few centuries may not correspond to that during the more distant past. These sorts of analogies work best when some cultural continuity exists, but even then many of the arguments are spurious. Archaeologists have gotten around this impasse in two ways: first, by looking at actual men and women—or at least their skeletons—to see how the work they did marked their bodies, and second, by looking at how different societies portrayed men and women in art.

SEDENTISM AND DECREASED STATUS: Some anthropologists have suggested that settling down invariably decreases women's status. In Botswana, the San, many of whom were mobile hunter-gatherers during the 1950s and 1960s, began settling in villages during the 1970s and 1980s. In many cases, the role of women within society shifted. When the group had been mobile, women had been just as important and active in the economic life of the tribe as the men. After settling down, they had to spend more time taking care of their new, permanent houses and were cut off from both the rest of the villages and opportunities to earn money. Men were the ones who went out with the cattle and "supported" the family. Certain chores—like drawing water—suddenly became low-status "women's work." Yet the great changes in San life over the last 50 years, like abandoning foraging to earn wages working in diamond mines or as cattle-herders, receiving government food subsidies, and attending government boarding schools, probably have at least as much to do with any change in women's status as giving up mobility does.

THE FEMALE INVENTION OF SEDENTISM: In fact, women may have chosen sedentism and were the ones who benefited from it the most. Mobility is not easy if you are trying to keep a toddler from playing with a cobra while you gather mongongo nuts. During interviews, San men and women have stressed different things when asked to talk about the difference between their old way of life as hunter-gatherers and their new, sedentary way of life. San men have reminisced about their hunting prowess and how much game there used to be in the good old days. San women, on the other hand, remember the hard work and often emphasize how much they prefer not having to move all the time, and getting to drink water whenever they want. Sedentism is generally associated with agriculture or intensive gathering activities, like harvesting marine resources. Although men do gather occasionally, plant processing is usually women's work. Once women came up with a way to stay in one place all the time merely by changing their plant-gathering activities, they probably jumped at it. Because of the importance of women's contribution to the entire group, men had to go along with this decision. Such a process seems to have happened in Sudan, among a group of foragers rather like the Natufians, who began cultivating wild sorghum more than 6,000 years ago.[39]

HEALTH STATISTICS AND THE INVENTION OF SEDENTISM IN WESTERN ASIA: So which scenario does the majority of the evidence support? Did sedentism make life easier or harder for women? A recent study, which considered the effects of the invention of agriculture on life expectancy in western Asia, shows that the ratio of males to females was higher during the Natufian than during the subsequent Neolithic period. The oversupply of men (and undersupply of women) at Natufian sites may be due to female infanticide.[40] Such practices suggest that this early sedentary society did not particularly value women. On the other hand, their status may have increased with the development of agriculture. Still, a study of 162 skeletons from Abu Hureyra shows that the transition to agriculture meant hard labor for most women. Female skeletons had deformed toe bones and well-developed upper arms, literally from keeping their nose to the grindstone, grinding grain all day, whereas men did not suffer such deformities and probably escaped this hard labor.[41]

INVENTING "WOMEN'S WORK" IN CALIFORNIA: The evidence from Abu Hureyra suggests that in some communities the dawn of village life also meant the dawn of "women's work" and increased division of labor according to gender. Among hunter-gatherers, great flexibility is the rule. Although men are often the hunters, and women often the gatherers, both share in the others' tasks occasionally, and no task is valued above the other. This is often not as true among villagers, who clearly differentiate gender roles. In Big Sur, California, the emergence of village life meant the emergence of "men's" and "women's" work, as seen from a combination of archaeology and historical accounts of these groups at the time of contact. During the early Holocene, it seems likely that both men and women did essentially the same type of work: gathering a wide variety of plants (including small seeds that they milled) and marine resources. Unlike in Sudan or the Levant, neither seeds nor marine resources were abundant (or nutritious) enough to enable these foragers to settle down. Over time, however, population pressure (and pressure on a few favored resources) encouraged foragers to give up on the easy life and settle down. Men developed a whole new toolkit of stone points and began hunting and fishing intensively. At the same time, women gave up gathering a wide variety of small seeds and focused on processing acorns, which are tasty and nutritious when leached and cooked. This is the same division of labor that existed in California at the time of contact. Just like in western Asia, settling down seems to have gone hand in hand with a new importance for the family. A sudden increase in valuable objects, like obsidian, otter pelts, and pendants, exchanged between families as part of marriage agreements probably results from a newfound importance for certain lineages.[42] What did all of this mean for the women and men actually living at Big Sur? At first, women's status may have increased as their acorn work allowed their families to survive hard times. Yet, their status would have decreased when marriages between lineages became important. Since a Californian women would not contribute to her family's wealth, but only to her husband's family's, raising girls became a waste of resources.

FEMALE AND MALE REPRESENTATION AT ÇATALHÖYÜK: Another way to learn about women's and men's status in early villages is to look at how the artists of the community portrayed them. The inhabitants of Çatal-höyük decorated their houses with murals, most of which showed wild animals and figures chasing after them. The beards of some of these figures show that they are male. Many of the animals are also male; stags and bulls with erect penises figure prominently. The molded animal heads that protrude from the surfaces of the houses are also usually male, displaying proud antlers. Women are not wholly without representation. Some wall paintings show them gathering plants, and female figurines can also be connected to plant use. One figurine even had a wild seed lodged in her back. Another figurine, depicting a heavily pregnant woman sitting on a throne made of two leopards, was found in a grain bin, perhaps uniting the world of plant manipulation and animal taming. Nevertheless, "the artistic evidence...points to a divided world, one dominated by males and their activities involving hunting and wild animals and the other, less frequently portrayed world involving women and plants."[43]

CONCLUSION

The origins of agriculture give us an opportunity to play the contingency game, to question what would have happened if history had taken a different course. Agricultural decisions made thousands of years before have influenced the way modern history has transpired. Yellow fever, which as we already saw, emerged along with slash-and-burn agriculture and irrigation practices in sub-Saharan Africa, prevented Europeans from ever developing populous colonies in the tropics. Children in Nigeria sang a nursery rhyme well into the 1980s that celebrates the deadly power

TIMELINE

	16,000 BCE	14,000 BCE	12,000 BCE	10,000 BCE	8,000 BCE	6,000 BCE	4,000 BCE	2,000 BCE	0 CE	Present
West Asia		Kebarans	Natufians	Agricultural Revolution, domestication of wheat, barley and rye Abu Hureyra 'Ain Mallaha Netiv Hagdud Mureybit	Domestication of sheep, goats, pigs and cattle. Ganj Dareh Origins of the village	Çatalhöyük, a complex Western Asian village	Spread of the horse? Spread of Indo-European languages to India and Iran?			
East Asia				Agricultural Revolution; Millet						
South Asia					Spread of agriculture from Western Asia		Spread of the horse? Spread of Indo-European languages to India and Iran?			Indian Ocean trade and the spread of agriculture
Central Asia					Spread of agriculture from Western Asia		Domestication of the horse			
Southeast Asia										Indian Ocean trade and the spread of agriculture
Australia and New Guinea					Origins of agriculture in New Guinea			Indonesian Australian and New Guinean contact		Complex foragers in Australia
Europe						Spread of agriculture from Western Asia Spread of Indo-European languages into Europe Complex foragers in Scandinavia	Spread of agriculture to Scandinavia			
North Africa						Saharan cattle herders	Desiccation of the Sahara	First domestic sorghum		
West Africa								Agricultural Revolution along the Niger River (yams) Spread of Bantu speakers		
East Africa								Sorghum and sedentary life in Sudan Agricultural Revolution in Ethiopia Spread of Bantu speakers	Indian Ocean trade and the spread of agriculture	
South Africa								Spread of Bantu speakers		
North America			Megafauna extinctions							
Meso-america			Megafauna extinctions	Foragers at Guilá Naquitez, Mexico	First domesticated squash, Guilá Naquitez, Mexico		First domesticated maize	First villages		Chumash chiefdoms in California
South America			Megafauna extinctions							

Last glacial maximum   Holocene   Younger Dryas Event

of yellow fever, the disease that turned West Africa into a graveyard for Belgian, British, French, and German colonists.[44]

Similarly, decisions not to develop farming have been disastrous in the long term for populations. British settlers easily displaced Australian aboriginal populations from prime areas because their villages could not defend themselves against imported weapons. Because few of the villages were sedentary, and no field system existed that the British could understand, they had no qualms in removing these native populations. New Guinea, an island that did develop agriculture, experienced a different form of colonialism. The farming highlands were never overrun by Europeans, or even by Indonesians. Today half of the island, Papua New Guinea, is independent, after the British, Germans, and Australians, all of whom tried to govern the island, pulled out in dismay.

Our views on agriculture have undergone a sea change in the last century. At the beginning of the twentieth century, when nearly 40 percent of Americans lived by farming, people thought agriculture and morality were inextricably linked. They viewed the simple figure of the farmer, tilling his own land, providing for his family, as uniquely noble. Today, according to the 2000 census, only 2.4 percent of Americans are employed in the agricultural sector. Perhaps as a result, we view the origins of agriculture as less important and less profound than our ancestors did. V. Gordon Childe, for example, the most prominent archaeologist of the twentieth century, described the Neolithic revolution as "the first revolution that transformed human economy gave man control over his own food supply."[45] Today, many archaeologists would scoff at such notions—agriculture is no longer seen as noble and unique; instead, archaeologists emphasize that it was simply one adaptation to the complex web of new resources available with the end of the Pleistocene. The true importance of agriculture lies between these two extremes. The revisionists are wrong when they downplay the significance of agriculture, stressing that it is no more "advanced" than fishing or no more important than the activities of harvester ants. Fishing did allow people to live in settled villages, use cowry shells for money, and even build totem poles, but it did not give them the ability to build empires. Like it or not, agriculture, albeit in its current highly industrialized form, is the foundation of our economy and shapes our experience of the world.

CHAPTER THREE

CHIEFS AND KINGS

OUTLINE

GETTING STARTED ON CHAPTER THREE: Cities, complex societies, and states all appeared for the first time about 5,500 years ago and have characterized human history ever since. Which forms of economic and political organization allowed for the emergence of the state? What forms did the state first take, and how and why did they differ across the globe? What was the relationship between cities, states, and governments in Mesopotamia, Egypt, and the Indus Valley? Did all three innovations characterize early complex societies in both the Americas and Eurasia?

In 2002, another archaeologist and I unearthed a series of rooms that looked to our twenty-first-century eyes like a workplace belonging to some low-level government workers. These offices were located at the city gate, inside the massive city walls of Tell Leilan, Syria. This agency used low partitions to separate these rooms, rather like today's corporate office cubicles. The artifacts we uncovered were not spectacular: mostly broken clay sealings (the third millennium BCE equivalent of customs forms) from containers of goods that merchants were trying to import and a lot of broken cups, which bore an uncanny resemblance to the tea-glasses currently popular in Syria. A lot has happened in the nearly 5,000 years since those Mesopotamian customs officials filled their days inter-rogating traders, examining their wares, and drinking pots of herbal tea (coffee and tea didn't make it to Mesopotamia until about 1000 CE, com-ing from East Africa and China, respectively). Yet the world they lived in parallels ours in many respects. Like ours, red tape regulated most spheres of life, and organizational ties—loyalty to the king or your place of employment—became important for the first time. A complex class system had emerged, characterized by an aristocracy who enjoyed privi-leges denied to the commoners, impoverished farmers who struggled to make ends meet, and even an urban underclass, who provided a vast re-serve of unskilled labor. In this sense, those bored custom officials repre-sent the beginning of our world.

To understand the transition to the first cities and states, we must first consider how states emerged from the various social forms that preceded them. Simply jumping from the origins of agriculture to the origins of cities does not give us enough information to figure out how and why people decided to build (and live in) large, dangerous cities—or even more importantly, to give up their own fields to live like slaves under a newly invented king. Before we can understand the state, we must begin by exploring the political structures and economic institutions of the societies that became states—often called chiefdoms. These structures included reciprocity—receiving goods and distributing them in turn—and surplus production—producing more than you need to survive. We will also explore specific chiefdoms in Mesopotamia, Mesoamerica, and the Peruvian coast. We will then consider how states and civilizations differ from other types of societies before discussing three early states: in Mesopotamia (3500–2500 BCE), Egypt (3200–2500 BCE), and the Andes (800–200 BCE).

After the Village, Before the State: Surplus, Chieftains, and Temples

CHIEFDOM POLITICS AND ECONOMICS

Societies followed several paths on their way to establishing states; there is no single pattern that defines all states.[1] Both Spain and the Tihuantin-suyu (the Incan name for their empire) were states in the sixteenth century, but they were obviously very different in outward appearance. We will discuss the predecessors of ancient states and how and why they developed certain techniques and institutions that encouraged the emergence of the state.

CHIEFDOMS AND SURPLUS PRODUCTION: Two of the revolutionary institutions that led to state development first appeared in chiefdoms: government—establishing a chief and nobles to make decisions for the community—and surplus production—producing more wheat, sheep, or pots than one family (or even a village) needed to survive. These two transformations are part of the "Rank Revolution,"[2] which means that for the first time, societies began to practice inherited difference. Suddenly, an accident of birth determined whether someone was a commoner or a noble and defined which paths were open to him or her. This meant that these new societies were no longer egalitarian, a striking difference from foraging bands and early agricultural villages. At the same time, the power of these new "ranks" was limited. Chiefs did not control the fields (or fisheries) of their followers. They may have had the birthright to wear beautiful feather coats, but they did not control the wealth of their society.

RECIPROCITY AND POLITICAL POWER: Chiefs gained power by manipulating the social rules of village societies, thus transforming these societies into something else. For example, chiefs in Polynesia and the Pacific Northwest gained political power through extravagant generosity. They could do this because the principle of reciprocity, wherein each gift demands one in return, operated in chiefdoms as well as villages. The more the chief gave away, the more he could expect to receive. The Eskimo proverb describes this policy's effects: "Gifts make slaves, as whips make dogs."[3] The principles of reciprocity also allowed chiefs to gain power by receiving gifts (or tribute) from people, storing them, and redistributing them in times of need. In Hawaii, the chief regularly collected taxes (in the form of taro root, the staple food) from the people and kept them in great storehouses. These storehouses gave chiefs the leverage they needed to

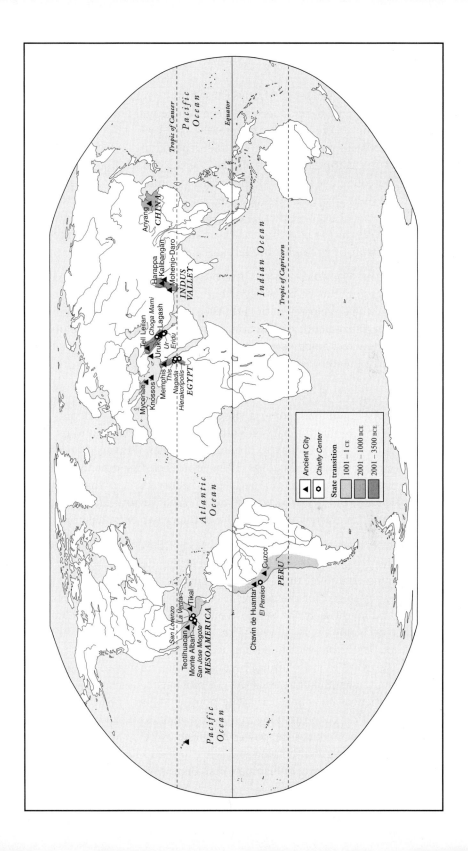

remain in power. As David Malo writes: "as the rat will not desert the pantry ... where he thinks food is, so the people will not desert the king while they think there is food in his storehouse."[4]

Yet chiefdoms remained fragile. Chiefs could quickly lose the support of their people during bad times and be overthrown. Unlike kings or dictators who could use armies against unruly commoners, or simply fire them from their government-supported jobs, chiefs had no way to force their followers to support them. As a result, if chiefs asked too much of their followers, their followers had no qualms about finding a new chief who was less uppity than his predecessor.

THE ECONOMIC FOUNDATIONS OF CHIEFDOMS: Whether they gained their power by giving gifts or receiving tribute, chiefdoms both relied on and encouraged the development of certain economic practices: surplus production, specialized craft production, and trade. Chiefdoms developed or employed new technologies such as irrigation agriculture and the plough to encourage greater yields. They began paying skilled craftworkers to make them feather capes, obsidian mirrors, or animal-shaped seals, and they encouraged traders to procure other special goods. Anthropological studies have shown that people in simple village societies work only hard enough to ensure that the harvest will feed them and their families. They find the idea of slaving away to produce more than they need or making things for other people ridiculous. They value their free time too much to put in more hours. Why did people suddenly eschew the notion of self-sufficiency and begin to work harder? What factors caused them to decide that extra grain was worth more to them than extra leisure?

THE ORIGINS OF SURPLUS PRODUCTION: Sometimes, simple households, despite the distaste or confusion they feel at the idea of producing more than they need, do just that in order to survive bad years. Excess grain may either be stored for the next year or given away to create debts. The first method might seem the best to avoid future starvation, but relying on stored crops is not always a good idea because they rot and can be eaten by vermin. On the other hand, giving away extra food can protect a household in the event of crop failure, because giving creates an obligation in the recipient, who must pay back the gift by offering hospitality when the giver experiences bad times. At some point, someone decided that they liked the political power that lay in this tiny bit of extra food. Chiefs arose when people decided that the power provided by being able to give away excess food outweighed the pain of having to spend long hours sweating in the garden (or freezing on fishing rafts). Because the more they worked, the more they could give away, and the more power they had, chiefs quickly adopted various tools that increased production. Hence, chiefdoms developed from villages and used many of the same organizing principles of such societies, like reciprocity. At the same time, the creation of chiefs meant a new social order beyond kin-based societies. No human society has ever existed in isolation, but chiefdoms traded more often and over greater distances than village societies. In some ways, chiefdoms were the very beginning of a "world system," where interactions between vastly different societies became necessary to the survival of many.

UBAID TEMPLES AND IRRIGATION AGRICULTURE

THE DEVELOPMENT OF IRRIGATION IN SOUTHERN MESOPOTAMIA: Irrigation is a tool that can dramatically increase agricultural yields. Although the first societies to use irrigation were not chiefdoms, once surplus production became desirable, irrigation fueled the development of chiefdoms in Mesopotamia, which were the prototypes for the first states. The earliest evidence for irrigation comes from Choga Mami, Iraq, and dates to 6000 BCE. Choga Mami lies at the very edge of the Mesopotamian desert, where it is slightly too dry to grow crops using rainfall alone. The original inhabitants of Choga Mami probably came from villages a few miles to the east, where more rainfall meant successful harvests. The immigrants at Choga Mami took advantage of the river Gangir, which flowed down from the Zagros Mountains, to provide their crops with the extra water that they needed to grow. At some point they even fashioned several small irrigation channels to take water directly to their fields.[5]

Although this probably seemed like a small innovation to the first villagers at Choga Mami, its impact was immense. It quickly allowed the settlement of the land adjacent to the Tigris and Euphrates rivers—Mesopotamia. This area was true desert, but the soil was incredibly fertile due to the build-up of silt from the river. After irrigation techniques had been invented, farmers could get much higher yields from these irrigated fields than they could from wheat or barley grown using rainfall alone—up to three times the amount. Immigrants from villages such as Choga Mami probably founded some of the first villages in southern Iraq, ancient Mesopotamia, along the Euphrates River, at the edge of the salt marshes that extend down to the Persian Gulf.

ERIDU AND THE FIRST TEMPLES: Although the earliest settlements uncovered in southern Iraq appear to be simple, egalitarian village societies, the picture soon shifts with the appearance of temples and the introduction of a system of "religious chiefdoms." One of the earliest sites in southern Iraq (ancient Mesopotamia) is Eridu, which, in later Mesopotamian history (once writing began), was the center of the god "of the deep"—the god of flowing waters and, no small coincidence, irrigation—Enki. Enki was a popular god in southern Mesopotamia, where life depends upon irrigation, so it is fitting that the first temple in the world was probably raised to him. In Eridu, archaeologists excavated a sequence of temples representing 1,500 years, a span of time referred to as the "Ubaid" (5500–4000 BCE), during which chiefdoms appeared and flourished in southern Mesopotamia. The earliest building was probably a simple house, but the next one had an altar with traces of burning on it, and later buildings had altars, offering tables, and niched architecture—all characteristics of later Mesopotamian temples. The latest temple, dating to immediately before the rise of cities and states, was built on a low platform—an ancestor to the later ziggurats, which served as the foundation for Mesopotamian temples. These ziggurats so impressed ancient peoples that a garbled account of the audacity it took to build one has survived until today, in the story of the tower of Babel.

THE APPEARANCE OF THE "TOWN":   As the citizens of Eridu enlarged and elaborated their temples, their settlement also grew until it became a town. At this point in southern Mesopotamia, there was a clear separation between towns 10 hectares (25 acres) in size, with a population of 1,000–2,000 people, and villages 1 hectare (2.5 acres) in size, with a population of 100–200 people. In general, the towns, like Ur and Eridu, contained clearly defined temples alongside the houses of their inhabitants, whereas villages did not.

The priest-chiefs who lived at Eridu or Ur seem to have translated their religious authority into political capital. In Mesopotamian mythology, Enki created humankind because the gods became tired of hard work and wanted slaves to dig the canals, sow the wheat, and make them bread. Hence, Mesopotamian temples were built along the same lines as human households. These temples owned agricultural land and herds that were used to feed the god and his human priests. Because Enki controlled the water necessary for life in the desert, the citizens of Eridu willingly worked for him—in the hopes that he would give them successful harvests in exchange. Because gods can eat only so much bread and mutton, the majority of the god's wheat and sheep were kept by the priests and redistributed to the people in times of need as a sign of Enki's generosity. Such a system gave the priests control over the three most important commodities in ancient Mesopotamia: land, labor, and grain.[6]

UBAID TRADING NETWORKS: The Ubaid towns bordered several environments: the fertile banks of the Euphrates, the harsh desert plateau, the wide marshes of southern Iraq, and the shallow Persian Gulf. Cores taken from the Gulf of Oman suggests that the shore of the Persian Gulf 7,000 years ago was not far from Eridu.[7] Perhaps as a result, there are clear trading networks between these southern Mesopotamian cities and the coast of Arabia. Scientific tests have proven that scraps of painted pottery found in Kuwait, Bahrain, Oman, and Saudi Arabia were made in southern Mesopotamia. It seems likely that traders from southern Mesopotamia traded grain, textiles, oil, and brightly colored pottery for pearls from the rich pearling grounds of the Arabian coast. These pearls may have been a sign of high rank for Ubaid chiefs. Thousands of fish bones from the Persian Gulf have been found piled on one of the temple altars, linking this long-distance trade with Eridu's home-grown religion. Although we have few signs of Ubaid imports—and even fewer signs of their exports—by the end of this period, people thousands of miles away in western Syria and Turkey were imitating Ubaid pottery. This suggests that irrigation, temples, and chiefdoms in southern Mesopotamia did not appear in isolation, but influenced people throughout southwest Asia.

TRADE AND ANCHOVETAS

Halfway across the world, 4,000 years after the rise of temples at Eridu, chiefdoms emerged along the Peruvian coast and in the Andes. Like Mesopotamia, little rain falls on the Peruvian coast, situated as it is between the high Andes on the one side and the Pacific on the other. Although there are perennial streams that allow for agriculture, the first chiefs who ruled over this coastal world and used their power to build a ceremonial center called El Paraíso did not rely on stored grain, but on tiny fish called anchovetas.

PERUVIAN FISHERIES: The cold ocean currents that run along the Peruvian coast support one of the richest fisheries in the world. Mollusks and large fish abound, but the true wealth of these waters lie in schools of tiny fish, called anchovetas. These fish can be dried and turned into protein-rich fishmeal. Modern estimates of the potential of these fish for supporting human societies are astonishing. As Brian Fagan writes, "judging from modern yields, if prehistoric coastal populations had lived at 60 percent of the carrying capacity of the fisheries and eaten nothing but small fish, the coast could have supported more than 6.5 million people."[8]

Unfortunately, during El Niño years, warm ocean currents do not provide enough nutrients for schools of anchoveta, causing them either to die off or go elsewhere. The inhabitants of the Peruvian coast may have initially begun drying and storing fish to get around these bad years. Peruvian chiefs probably exploited these conditions by saving more fish meal than anyone else and giving it out in hard times, thus placing the rest of the starving inhabitants in their debt.

IRRIGATION AND EL PARAÍSO: The ancient Peruvians may have eaten a lot of anchovetas, but they also had gardens watered by streams from the Andes. By 1800 BCE, the chiefs had found a new source of social power: large irrigation projects. Irrigation allowed them to grow a few food crops, like beans, gourds, and squash, as well as cotton, which they turned into elaborate textiles. Due to the extremely dry conditions prevalent on the Peruvian coast, beautifully woven and dyed clothing dating to as long ago as 4500 BCE has been found. At about 1800 BCE, coincident with the introduction of large-scale irrigation works, Peruvian weavers came up with a new type of spindle that allowed them to produce finer cloth. Throughout history, cloth has symbolized wealth or high rank. In the Roman Empire, for example, rich purple dyed cloth (from Tyre) was so costly and symbolic that the amount of purple on one's robe was dictated by status and enforced by law. Only the emperor was wealthy enough (and powerful enough) to afford an entire robe. It seems likely that Peruvian chiefs used ornate cotton mantles, shirts, and turbans to emphasize their rank in the same way.

TRADE AND THE DEVELOPMENT OF EL PARAÍSO: Another source of wealth, similarly controlled by the chiefs, came from trade. The appearance of El Paraíso on the coast coincides with the appearance of other similar ceremonial centers in the interior. At the same time, trade between the coast and the highlands increased. Crops like potatoes, *oca*, and *ullucu* that can be grown only in the mountains have been found along the coast. The coastal elites may well have traded salt, fish, and seaweed (which contains iodine—an element which was absent from the high-starch diet of the mountain dwellers) for these foods. Other commodities, like brightly colored cloth or alpaca wool sweaters, might also have been traded between the highlands and the lowlands. These trade goods were luxury items, probably given away at huge feasts, which further cemented the high rank of the chief who organized them.

CHILDREN OF EARTH AND SKY: THE OLMECS AND THE ZAPOTECS

In Mesoamerica, several competing chiefdoms arose at roughly the same time, across a wide variety of ecological zones—from mountain deserts to coastal rain forests. The chiefs who ruled them were born into chiefly lineages, but gained power and wealth during their lifetimes. These chiefdoms were funded both by intensifying agriculture and sponsoring trade in exotic goods and the manufacture of other prestige goods. We will focus on two of these chiefdoms: the Zapotec, who lived in the mountains of Oaxaca, and the Olmec, who lived near the Tabasco coast.

THE DEVELOPMENT OF INHERITED RANK: One of the main differences between chiefdoms and kinship societies is that in chiefdoms rank is inherited and is a major structuring force in society. In kinship-based societies, an individual can achieve high rank and his neighbors can esteem him, but he cannot pass that rank on to his son. His son must earn it himself. In chiefdoms, by contrast, although each chief still has to compete against his brothers and cousins to become the paramount chief, he does not have to build up support for the institution of chiefdom. The commoners will sow and harvest crops for the temple and give some of their surplus food to the chiefly storehouses because their mothers and fathers did so for the chief's father, their grandparents for his grandfather, and so on. As a result, each ambitious leader-to-be does not have to convince each skeptical villager to support him. That support is already built into their society.

INHERITED RANK IN OAXACA: After studying hundreds of graves in Oaxaca, Mexico, Kent Flannery and Joyce Marcus believe they have evidence for inherited rank and the beginnings of chiefdoms among the Zapotec. The Zapotec's two most important gods are Earth and Sky, who are symbolized by an earthquake icon and a lightning icon, respectively. At some point, it seems that certain Zapotec families began tracing their ancestors back to Earth and Sky. Males of chiefly lineage were buried with vessels painted with either Earth or Sky motifs. Even infants could be buried with such vases. Because there is nothing a baby could have done to earn the status to which these specially made vessels attest, it seems likely that they were born with high rank. At the same time, not all adult men buried with Earth or Sky jars became chiefs. Each chief still had to consolidate his power by fighting all the contenders for the title—the other children born of "noble lineage." Chiefs were even buried in ways that showed their authority; their bodies were tightly bound together with the knees pulled up to the chest, so they remained in a sitting position, sometimes perched on a stool. This was in contrast to the prostrate position used in the vast majority of Zapotec burials. In later Mesoamerica, kings did not touch the ground, but always sat on stools or benches so they were higher than the commoners. The burials of Zapotec chiefs reveal the emerging ideology of the ruling class.[9]

THE CHIEFS OF SAN JOSÉ MOGOTE: The Zapotec chief lived in a town now called San José Mogote. Unlike the other villages in the valley, San José Mogote contained two simple temples on pyramidal platforms made of limestone brought from elsewhere in the valley. This town was much larger than the villages around it and may have had a population of 1,000 people. Before the appearance of inherited status and chiefly rank, the site only had a population of about 100–200. After chiefdoms appeared at San José Mogote, immigration also increased to the valley of Oaxaca as a whole—tripling the population until it reached at least 2,000. All of these people could live in a small part of the Oaxaca valley because of new innovations in agriculture, which led to larger harvests and surpluses that could both feed the immigrants and support chiefly building projects. Prior to the arrival of the chief, only the permanently humid bottomland of the valley was farmed. After the chief assumed leadership of the community, the Zapotec began using two forms of irrigation. The first was jar irrigation, which simply entailed digging wells and pouring jars of water on the field; the second involved digging small ditches, which could be used both to channel streams into the fields as well as drain away excess water.[10]

ZAPOTEC TRADE: The Zapotec chiefs used high status items, which they obtained from local craftsmen and from trade with the rest of Mesoamerica, to exhibit their newly forged authority. Magnetite from the Gulf Coast, which could be made into mirrors or just admired for the way it caught the light, was restricted to chiefs and nobles. Shell, from both the Pacific and the Gulf, along with jade, were also signs of wealth and power—as well as evidence for trade contacts with a wide area of Mexico.

ZAPOTEC WAR AND DIPLOMACY: Trade was not the only way that the Zapotecs interacted with the other emergent chiefdoms of Mesoamerica. Much like later states, the Zapotecs waged wars and made peace according to their own interests. Although Mesoamerican chiefdoms did not have enough power (or the ability to control a standing army) to wage long wars or to administer conquered lands, they contented themselves by raiding other chiefdoms. Generally these raids would end when the winners burnt the temple of the rival chiefdom, killed its chief, and took

FIGURE 4

Danzante from San José Mogote that was found in a ceremonial building complex atop Mound 1. Monument 3, San José Mogote.

a few prisoners. A threshold stone between two monumental buildings at San José Mogote illustrates one of the fates that befell warriors unlucky enough to become POWs in the days before the Geneva Convention.[11] The man pictured on the slab lies awkwardly on his back, with his chest torn open and his heart removed. Streams of blood flow from this gaping wound, and the closed eyes, open mouth, and pained expression on his face show that the dead man suffered terribly in his last moments.

In other situations, however, Zapotec chiefdoms sealed strategic alliances with diplomatic marriages. The burial of a high-status woman (perhaps the daughter of San José Mogote's chief) found 5 kilometers away, at a hamlet where salt can be retrieved from saline springs, illustrates this phenomenon. Chiefdoms also interacted peacefully through competitive feasting, trying to surpass each other's hospitality. This gave both chiefdoms an incentive to amass all the luxury foods and fancy clothes that they could find, strengthening their political system as well as their economies.

OLMEC CHIEFDOMS: The Olmec, who lived on the tropical Veracruz coast, were trading, raiding, and marriage partners of the Zapotec. Like the Zapotec, the Olmec chiefdoms seem to have appeared around 1150 BCE and to have established their capital at a settlement now called San Lorenzo. It seems that the conch shell trumpets, stingray spines, and turtle-shell drums that have been found at San José Mogote came from San Lorenzo. Although no graves from this period have been excavated, some Olmec pottery, like that of the Zapotec, was decorated with symbols of Earth and Sky (earthquake and lighting). Unlike the Zapotec, with their pyramidal temples or rich graves, however, the labor and wealth controlled by the Olmec went into the manufacture of huge basalt sculptures. Skilled craftsmen carved giant heads, sometimes human (perhaps portraits of Olmec chiefs), sometimes half human and half jaguar, out of basalt blocks weighing several tons. The Olmec chiefs were able to persuade their followers to drag these heavy boulders for 60 kilometers—which took some convincing.

The Olmec chiefs rewarded their followers with corn, squash, and beans grown on raised fields (*chinampas*), a technique that, like irrigation, greatly increases agricultural yields. Also like irrigation, chinampas cannot be built or maintained by one person, but only by a community. The Olmec chiefs probably supervised their construction and upkeep. By 800 BCE, Olmec chiefdoms had become rich enough to supply both the workers and the corn needed to feed them in order to build La Venta, a ceremonial center on an island in the midst of a swamp. A rectangular plaza, a series of long, low mounds, and larger terraced mounds dominate the site. Huge stone sculptures and altars are everywhere—some of which weighed more than 40 tons. A throne from La Venta shows a chief connected to his ancestors on either side by umbilical cords, giving definite proof for the importance of inherited rank in this society.

The Zapotec and Olmec chiefdoms provide a good conclusion to this section for three reasons. First, Mesoamerican chiefdoms were probably some of the richest, most powerful chiefdoms ever to exist; as a result, by examining them we can find the dividing line between a chiefdom and a state. Second, these same chiefdoms invented writing and became either the first, or at least the forerunners of, Mesoamerican states and civilization. Third, like the other early chiefdoms we have examined, Mesoamerican chiefdoms did not emerge alone. They were "sister cultures"; their awareness of each other and mutual competition made them stronger and richer faster.

THE POLITICAL AND ECONOMIC FOUNDATIONS OF ARCHAIC STATES

What is the difference between chiefdoms and states? Scholars often describe state societies as "complex societies" or civilizations—yet chiefdoms like the Olmec or the Zapotec are far from being simple. Furthermore, their art, religion, and early writing are recognizable precursors to those of later Mesoamerican civilizations. In general, when historians or anthropologists talk about a "state" or a "state society," they are referring to a form of political organization that differs from a chiefdom (as well as all other forms of political organization) because it requires both political and economic hierarchies. Hierarchy, also referred to as "stratification," is what allows a state to exist, and it can explain all the other attributes that we commonly see in states, such as monumental public buildings, administration, writing, the invention of sciences, the formation of professions within a complex division of labor, and so on. "State" is not just a synonym for government; government is only one aspect of the state—the institution that helps to maintain it and its personnel.

WHAT MAKES A STATE?

DEFINITIONS OF THE STATE AND POLITICAL STRATIFICATION: To be a state, a political system must have four or more tiers of administrators.[12] A simple foraging band, like the San, will have at most one tier—a leader who may decide when they will make and break camp, but who is often ignored by the rest of the band. Simple chiefdoms usually have two tiers, the chief and the other members of the chiefly lineage. A state, however, must have four or more levels (a king, a powerful circle of political advisers, their underlings, etc., down to the poor man who has to sit all day in the gatekeeper's office). States also have four levels of settlements: a capital city that is probably several times the size of the next largest settlement, a second tier of large towns, usually equidistant from the city, a third tier of small towns (centered on the large ones), and a fourth tier of villages, usually administered by the small towns.

ECONOMIC STRATIFICATION: States must also be stratified economically, not just ranked socially. In the other social arrangements we have examined, people have been pretty equal in economic terms. In chiefdoms, chiefs might have a lot of prestige, but usually they do not have that much economic power over their followers. The chief may be able to encourage surplus production, and may even tax his followers (who pay begrudgingly, because he is, after all, the great-great-great grandson of Sky) and get them to build a few community projects, like ceremonial centers or carved heads, but he cannot force them to do anything. This is because the commoners living in chiefdoms own and farm their own land. They are not dependent on the king for their survival. They may enjoy his feasts, and the handouts from his storehouse may be very welcome in times of famine, but on the whole the commoners do not need the chief, whereas he needs them. Chiefs gain power through bribery, but they cannot coerce their followers. If the chief goes too far, the commoners will overthrow him by supporting another member of the chiefly lineage.

THE ORIGINS OF INEQUALITY OF OPPORTUNITY: With the arrival of the state, this shifts dramatically. Suddenly, not everyone has equality of opportunity (or equal access to the means of production). Instead, a limited number of people have the ability to become rich, while the majority of the population must either work for them, or starve. All the inhabitants of the first cities did *not* own fields. Rather the temple, the king, and a class of nobles owned the majority of the land. Although some small landowners continued to exist, they had to pay taxes to the king in the city, or risk having their property confiscated by the state. The state used all the food it grew and collected taxes to pay people to be full-time professionals, like potters, cooks, or bureaucrats. These experts no longer had to spend all day in the fields, only devoting their spare hours to pursuing their passion for weaving or soldiering. The administrators who sat at the city gate at Tell Leilan did not tend their own fields or herd their own sheep. Instead, every month, the state gave them rations: so many liters of grain, so many liters of beer, and occasionally a woolen tunic. The state created two types of hierarchies: administrative hierarchies, encompassing functionaries from kings to clerks, and economic hierarchies, encompassing everyone from nobles to slaves.

PRIMARY AND SECONDARY STATES: Scholars usually write about two types of states: primary states, which are the world's first states, all of which emerged independently, and secondary states, newer states which emerged only after contact with previously existing states. Primary states are extraordinarily rare; there are only six places where states probably emerged independently: Mesopotamia, Egypt, the Indus Valley, China, Mesoamerica, and Peru. Secondary states encompass every other part of the world, including all of Europe, central Asia, southeast Asia, sub-Saharan Africa, North America, and Polynesia. They have emerged over a long time period, beginning with city-states, like Tell Leilan, which came to prominence at about 2600 BCE in northern Mesopotamia, and continuing with states like Papua New Guinea, where isolated groups of foragers and farmers are still being integrated into a new country (founded in 1975). But even the primary states in western Asia and northern Africa—Mesopotamia, Egypt, and the Indus Valley—all established civilizations and invented writing within a few centuries of each other. Moreover, they traded with and influenced one another. The state was such a powerful and revolutionary social type that it changed the way societies, sometimes even distant ones, worked. The appearance of the state marked a major change in the history of humankind.

THEORIES OF THE ORIGINS OF THE STATE

Why do states emerge? One way of exploring this question is considering actual cases where chiefdoms have evolved into states. Two hundred years ago, Europeans who had established trading contacts with native Hawaiians witnessed the transformation of Hawaiian society from competing chiefdoms to a unified state. Looking at how a chiefdom developed into a state on Hawaii helps us imagine what could have spurred a similar system to transform itself in antiquity.

THE TRANSITION FROM CHIEFDOM AND STATE IN HAWAII: When Captain Cook arrived in Hawaii in 1778, rival chiefs ruled each of the large islands and fought to control the smaller islands. In 1782, the chief of the big island of Hawaii died and three of his descendants started competing for his title. One of them, Kamehameha, managed to beat his rivals by trading with the newly arrived Europeans, acquiring guns, cannons, and "even two European officers who served as his gunners and strategists between 1789 and 1790."[13] Not only did Kamehameha seize control of the whole island of Hawaii, he also conquered all the other islands, except for Kaua'i—which surrendered in 1810 when its chief realized that it could not maintain its independence. European guns and ships may have given Kamehameha the advantage, but what really turned Hawaii into a state were the institutions he invented to govern his new territory. These included a new emphasis on economic intensification and the invention of professions: Kamehameha created a standing army that he fed by intensifying agriculture on the islands under his control.

"GREAT MAN" OR ACTION THEORY AT MONTÉ ALBAN: Some archaeologists believe that a single individual, a great king like Kamehameha, is necessary to transform a chiefdom into a state. This is often referred to as the "great man" theory, or "action" theory. An example of this is the transformation of the Zapotec chiefdom, described in the previous section, into a state. This probably occurred when a great chief, in response to escalated raiding, convinced the inhabitants of several towns and villages to form a city, Monté Alban, on a previously uninhabited mountain peak. To feed all the people in this new city, the Zapotec king had to organize a new agricultural system and distribute its produce. Because no one had lived there before, he may have claimed ownership of all of the land and then allotted it to members of chiefly families from different towns and villages that now lived in Monté Alban. Its creation brought together craftsmen from throughout the valley: the jade-workers, shell-carvers, mirror-shiners, and so on. To optimize their talent, the new king probably instituted a ration system to support the artisans. This would have been extended to administrators, who were suddenly necessary to administer the artisans and farmers that fueled the system. At the same time, the scattered hieroglyphics used previously at San Jose Mogoté evolved into a coherent writing system, one that only gave the elites more power. Although the many changes in the political system at Monté Alban did not occur under one individual, an especially charismatic king probably supplied the necessary push, which set into motion this elaborate chain of events.

WARFARE AND THE ORIGINS OF THE STATE: Another theory argues that, as in Hawaii, warfare can stimulate the creation of the state. The areas where early states arose, although located in different parts of the world, all have one thing in common: restricted access to agricultural land. In the forests of the Amazon, or northern Europe, the amount of good land for an early farmer was almost infinite. In such an environment, people could avoid conflict by moving away from their neighbors. As a result, at the time of European contact, villages covered almost the entire Amazon, but they were all located quite far from their neighbors. In places where states developed—Mesopotamia, the Peruvian Coast, Egypt, Mesoamerica, the Indus Valley, and China—the amount of land was limited by mountains, deserts, or oceans. As a result, villagers could not just move away to avoid aggression, because all of the land had been taken. When villagers lost a war, they had to stay where they were. If their conquerors let them live, they did so as a conquered, subservient village. Population growth accelerated the process; the more people there were, the less new agricultural land, the greater the likelihood of warfare. And so simple villages became chiefdoms and warring chiefdoms became states. The natural outgrowth of these new states was the development of empires, which generally emerged a few centuries (or millennia) after the state.

METALLURGY AND THE STATE: Still others suggest that the introduction of new technology is behind the sudden appearance of the state. If guns and cannons led to a Hawaiian state, then copper and bronze metallurgy created the weapons that led to the first states in Mesopotamia, Egypt, and the Indus Valley. V. Gordon Childe was the first modern archaeologist to write about the origins of cities and states. He portrayed this as the second great transformation in human history (after the invention of agriculture) and coined the term, "the urban revolution" to describe it. Childe believed that the invention of metallurgy was the critical first step to a class system, full-time professions, and powerful kings. Why? Copper and bronze weapons give warriors an edge over their less technologically advanced enemies. These metals are harder than stone, cut better, and last longer. As Childe says: "it does not matter much if a flint knife breaks in skinning an animal. It is a much more serious matter if the accident happens in hand-to-hand combat with an enemy."[14] Armed with bronze weapons, these warriors would have vanquished their rivals in battle after battle. Because the river valleys of Mesopotamia, Egypt, and the Indus were filling up with people, the inhabitants of the defeated villages could not have escaped, but would have become the slaves of the bronze-wielding elite instead. Unlike pottery making, stone carving, or most other crafts, being a blacksmith is a full-time job. Ancient coppersmiths would have had to be supported by farmers. Because what they did was so important, these metalworkers would probably have had a higher status than the farmers who fed them. Unfortunately, Childe's hypothesis has been challenged on several grounds. Part-time blacksmiths have existed in nonstate societies—in African villages and among Central Asian pastoralists, for example. Similarly, metallurgy first appeared in Europe, in societies much less complex than the world's first states.

HYDRAULIC SOCIETIES: Another theory of the origins of the state also relies on the importance of technology, but is concerned with the development of irrigation, not warfare in times of scarcity. Most early states arose in river valleys and relied on irrigation for their existence (even Hawaii used irrigation for its taro and manioc plantations). According to this theory, although just one person can sow some seeds, wait for rain, and then use the resulting harvest, it takes a state to build and maintain a large irrigation system. To use irrigation efficiently, it is necessary to dig networks of canals that distribute water across a plain. The work does not end after they are built, as silt from the river blocks canals each year, forcing farmers to clear and straighten them. Building and maintaining an entire network of canals is a lot of work, far more work than any one farmer—or even a family or a village—would be able to supply. As a result, a leader is necessary to direct the work crew, to organize the labor and inspire the villagers and towns-people to share in a common sense of responsibility. Some historians have proposed that the leaders of the work crew capitalized on this power by becoming the ruling elite of early "hydraulic" (irrigation-based) societies.[15] Organizing large work groups for monumental architecture or military service was a natural outgrowth of organizing irrigation work crews. Archaeological work, however, has shown that the irrigation systems that existed at the time of the first states were very simple. Intricate irrigation systems appeared long after the state; families, or chiefdoms could main-tain the earliest of these systems.[16]

SPECIFIC AND MULTICAUSAL THEORIES: Some scholars eschew the example of Hawaii and suggest that states rose in response to local circumstances. Others caution us to be careful about seeing any one cause behind some-thing as complex and varied as the state. Instead, they argue that the state came about because of a combination of many reasons. Separating out the one development that absolutely led to the state in any location is impossi-ble. When comparing Mesoamerica and Mesopotamia, for example, Robert Adams stressed that irrigation agriculture, increased warfare, and resource variability—being able to use resources from the mountains, valleys, des-erts, riversides, and marshes—all contributed to the rise of the state.[17] Some communities became more powerful than others because they had better land, more people, or better soldiers. They came out on top and were able to force their poorer and weaker neighbors to do their bidding. The victorious communities developed into religious centers (after all, the gods gave them their victories), as well as centers for artisans and innovations. As a result, powerful chiefdoms transformed themselves into the world's first states.

INVENTING THE STATE

MESOPOTAMIA: SO YOU WANT TO BUILD A CITY

The Epic of Gilgamesh, often considered the world's first work of litera-ture, begins and ends with a description of the city of Uruk, the capital city of the ancient and powerful Mesopotamian king Gilgamesh:

Go up, pace out the walls of Uruk,
Study the foundation terrace and examine the brickwork.
Is not its masonry of kiln-fired brick?
And did not seven masters lay its foundations?
One square mile of city, one square mile of gardens,
One square mile of clay pits, a half square mile of Ishtar's dwelling
Three and a half square miles is the measure of Uruk![18]

Uruk, which at its height in the mid-third millennium BCE, was 400 hectares (988 acres) and may have had a population of 80,000 inhabitants, was the world's first city and the capital of the world's first state. The city was at the center of Mesopotamian civilization; archaeological surveys have shows that more than 80 percent of the early Mesopotamian (Sumerian) population lived in urban centers, defined here by cities larger than 10 hectares (and with a population of about 2,000 people). Using a similar definition, the United States in the twenty-first century CE is 5 percent less urban than Mesopotamia was in the third millennium BCE.[19]

THE MESOPOTAMIAN URBAN REVOLUTION: Archaeologists have spoken of the rise of states as an "urban revolution," emphasizing the importance of cities in civilizations, although few other ancient civilizations relied on urbanism as heavily as Mesopotamia. But what makes a settlement a city—in contrast to a village or town? Part of it, of course, is size. Cities covered a larger area and had far more people than agricultural villages did. But the people that the city brings together, and the institutions that govern them, are far more important to an urban identity. Ancient Uruk was not made up of uniform houses; instead, it housed elaborate temples, large residences, and small houses arranged in neighborhoods. In later Mesopotamian cities, these neighborhoods often had special characters—rather like cities in the Middle East today, where all the goldsmiths have their shops (and houses) in one quarter, all the weavers in another, and so forth. Uruk was not simply a village of farmers; it was a city of artisans, soldiers, bureaucrats, nobles, and priests, as well as farmers (someone had to feed the specialized professionals). To distribute food and raw materials, train skilled workers, and keep everyone from killing each other, Uruk needed a basic form of government.

URUK AND THE RISE OF THE TEMPLE: No "palace" has been excavated in Uruk. Indeed, the first palace (as well as the first reference to a definite king) appears at approximately 2600 BCE, 900 years after Uruk became a state, at approximately 3500 BCE. In the fourth millennium, Uruk was full of monumental buildings, but as the *Epic of Gilgamesh* suggests with its emphasis on

"Ishtar's dwelling," those buildings were temples. Yet these temples are a far cry from the simple niched buildings in Eridu. Instead, the walled "Eanna Complex" (a Sumerian name meaning House of Heaven), enclosed several separate shrines. These were built of mud brick, but decorated with thousands of small baked clay cones whose red, white, and black heads were set in a variety of designs. Another temple area was placed half a kilometer west of Eanna in the Anu complex (Anu was the sky god of the Mesopotamian pantheon). Here the "White Temple," a building coated with gleaming gypsum plaster, towered 50 feet above Uruk.

The early rulers of Uruk probably styled themselves as high priests of the gods and used the existing prestige of the temples to cement their authority. The Warka vase, which was found at Uruk, dates from approximately 3000 BCE and depicts a procession of people bringing offerings to the goddess Inanna, with the city-ruler surveying the work. This vase provides evidence that early rulers used their devotion to the goddess to harness wealth and power, which they used to govern in her name, much the way that earlier chiefs probably came to power in Eridu.

This does not mean, however, that the priests and temples were all-powerful. Because the oldest Sumerian records come from temples, where they were used to keep track of commodities, scholars thought that the temple owned all the land and animals in Sumer. Today, this theory is regarded as naive, and scholars suggest that the temple was not all-powerful. Instead, individuals and families probably also participated in the economic life of the community.

To fund their building projects (as well as to feed themselves), the rulers of Uruk taxed people from the immediate countryside as well. In fact, Uruk may have ruled an even larger state, one whose borders are not entirely clear. Mesopotamia has gone through many periods where a strong, centralized state united the countryside, and other periods where this area was politically fragmented (ruled by city-states), although culturally unified. The city of Uruk grew as people left small nearby villages to move into this metropolis. At the same time, the need for more food probably motivated the rulers of Uruk (and those of other large cities) to organize the countryside.

In the 1950s, a book titled *History Begins at Sumer* was a popular history best seller. The title records an indisputable fact: although states have emerged many places, they emerged first in Mesopotamia. The same goes for cities, writing, mathematics, and much more. As a result, archaeologists working all over the world have looked at the example of Mesopotamia to explore the origins of states. Before Uruk arose, the world was inhabited by many groups of simple and complex hunter-gatherers, a fair number of agricultural villages, and a few chiefdoms. Soon afterward, the map of North Africa, the Mediterranean, and south and west Asia had to be reconfigured to make way for new states. What caused the rise of the state in southern Mesopotamia during the mid-fourth millennium BCE?

INTERACTION, CLIMATE CHANGE AND THE RISE OF THE STATE IN MESOPOTAMIA: Southern Mesopotamia was situated in a unique, varied environment. People who lived there could take advantage of the water-logged soils next to the rivers for date and poplar orchards, the more distant irrigable land for wheat and barley, the canals and marshes for fish and fowl, the desert for herding sheep and goats, and the Persian Gulf for marine resources. As a result, well before the state emerged, different groups had begun to live and specialize in different environments: villagers tended orchards and fields, nomads kept flocks of sheep and goats, and marsh dwellers fished and hunted. It appears that changing rainfall patterns, which made southern Mesopotamia drier during the fourth millennium BCE, encouraged both trade and storage.[20] Bad years for fishermen were not necessarily bad years for farmers. The different groups began to trade among themselves, creating dependent relationships.

From the beginning, trading was more than simply exchanging different types of food to prevent starvation. Nomads traveled far and traded widely; they did not just supply villagers with occasional cuts of mutton, but probably also passed along obsidian and silver from Anatolia and lapis lazuli and carnelian—semiprecious stones from Afghanistan and western India. Similarly, fishermen may have brought back copper from long-distance fishing trips to Oman in the Persian Gulf. Emerging towns (and later cities) became the center of this trade, as well as cult centers that could unite the diverse populations of southern Mesopotamia. As cities became larger and people came into contact with one another more frequently, innovations multiplied. Trade systems were administered and controlled so that suitable quantities of everything from copper to fired bricks to tasty perch were available at all times. Because Mesopotamia contained few natural resources (except for mud, reeds, sheep, grain, and poplar trees—all of which it had in abundance), they traded manufactured products; for instance, they might exchange the latest fashion in linen evening gowns and the precursors to Persian rugs for the raw metals and precious stones of the nearby mountains. Producing these trade products forced the Mesopotamians to establish and administer textile workshops, as well as to distribute the raw materials that came from the nearby mountains.[21] Thus, civilization in Mesopotamia developed through a combination of climatic, geographic, and cultural factors.

EGYPT: AN EMPIRE OF VILLAGES

Although civilization emerged in Egypt only two hundred years after Uruk began its long rule over the southern Euphrates, and after centuries of trade ties with Mesopotamia, the two cultures could not be more different. If Mesopotamia was a heartland of cities, an area where urbanism and civilization were synonyms, then Egypt was an empire of villages. Not until Alexandria was founded in 332 BCE, did Egypt have a city that in any way rivaled large Mesopotamian cities like Uruk, Ur, Nineveh, and Babylon. In Mesopotamia, loosely united city-states were the rule, and periods of empire were the exception, at least during the second and third millennia BCE. In Egypt, on the other hand, unity was the rule, celebrated as the natural order, while disunity meant that chaos, a greatly feared natural force, was winning its struggle against the gods.

EGYPTIAN POLITICAL AND GEOGRAPHIC UNITY: Egyptian political unity was encouraged by geography because "the Nile Valley is really only an extremely elongated oasis."[22] From the Sudan to Cairo, the river cuts a narrow strip of agricultural land through one of the world's harshest deserts. North of Cairo, the Nile breaks up into numerous branches that flow into the Mediterranean and create the fertile delta. Traditionally, Egypt was divided into two parts, the delta (Lower Egypt) and the river valley (Upper Egypt). Egypt was a long, narrow country, but because it encompassed a navigable river, officials could sail from Memphis at one end of the kingdom to Elephantine, at the other, in just about a week. Unlike Mesopotamia, where the Euphrates and Tigris often switched courses, leading to disastrous floods, making irrigation a risky business, and causing some areas to be much more fertile than others, Egyptian land along the Nile was uniformly fertile. As a result, it offered few natural opportunities for cities to develop and encouraged small regular villages throughout its course. It also offered few opportunities for exchange, because each village could expect a similar good or bad yield each year. But there is definite evidence of more long-term exchange in Egypt. Copper, gold, and turquoise were mined in the eastern desert—or possibly acquired by trade. Lapis lazuli came from Afghanistan, via Mesopotamia. The Egyptians also imported cylinder seals (administrative tools made of carved stone) from Mesopotamia. The popularity of certain Mesopotamian motifs in Egyptian art, as well as the use of decorated clay cones (like those used at the Eanna complex in Uruk) also illustrates exchange between these areas.

PREDYNASTIC CHIEFDOMS: According to later tradition, Egypt became a state when King Narmer or the Scorpion King unified it at approximately 3200 BCE. Before this time, around 3500 BCE, three chiefdoms had existed in Upper Egypt, centered on the towns of Naqada, Hierakonpolis, and This. The delta was organized more simply than Upper Egypt, although there is evidence of chiefly burials at the site of El-Omari. The Upper Egyptian chiefdoms were ruled by charismatic chiefs who cemented their power by controlling both the abundant harvests of the Nile valley and the long distance trade in luxury goods. Paintings from the tomb of a chief at Hierakonpolis, perhaps the center of united Upper Egypt, depicted scenes that resemble later Egyptian artistic propaganda, but are painted in a foreign style. These scenes show the chief smiting cowering captives with a mace and the chief standing under an awning celebrating a festival, which resembles later royal jubilee ceremonies.[23] When this chiefdom was powerful enough to conquer and to administer the ecologically distinct delta (Lower Egypt), the Egyptian state was born. The Narmer palette, an engraved slate slab, provides an artistic rendering of the union between Upper and Lower Egypt (Figure 5). On one side, King Narmer wears the white crown of Upper Egypt and is shown killing a captive, while on the other he wears the red crown of Lower Egypt and calmly inspects decapitated enemies from the delta.

DIVINE KINGSHIP: As befits a state that grew out of chieftainships that emphasized the importance of the ruler, Egypt established a unique form of divine kingship early in its history. Unlike Mesopotamia, where the first kings were just priests of the city-god or goddess, in Egypt, the king was a god. A complex ideology quickly grew up around the notion of the pharaoh, which helped the kings to consolidate and maintain their position as rulers. As Brian Fagan comments, "In Egypt, the terms *father, king* and *god* were metaphors for one another and for a form of political power based on the inequality considered part of a natural order established by the gods at the time of creation."[24] The king preserved order, which allowed the world to function the way it should. In Old Kingdom Egypt, there were no words for government or state except for the word for *king*. As a result, the entire state was centered on the ruler. In Egypt, people conceived of history as a list of kings—a view that remains popular with some today. One king list, written in the New Kingdom (1500–1000 BCE), included the names of all the kings of Egypt for the previous 2,000 years. Before the names of the historical kings, the list contained the names of several spirits and before these names, a list of gods. Although this Egyptian

FIGURE 5

Narmer, unifier of Egypt, prepares to sacrifice an enemy. He wears the crown of Upper Egypt, and the falcon representing the god Horus oversees his actions in this relief carving on a votive tablet. Two fallen enemies lie at the bottom of the tablet. Narmer Palette.

historical document seems far from accurate to us now, it powerfully combines mythology and history into an ideological statement. For the Egyptians, their country had always been unified, and the rule of the kings today was a continuation of the rule of the gods. From this curious papyrus, "the ancient scribe could have known the age of the world since the time of the first creator god and he would have seen how the kings of the past and their great monuments fitted within this majestic scheme."[25]

PYRAMIDS AND KINGSHIP:　It is not much of a surprise, then, that although Egypt had an elaborate bureaucracy from the earliest period, the most important projects glorified the king. In many people's minds, pyramids are synonymous with Egypt. The Egyptian kings began to build pyramids during the reign of Djoser (2630–2611 BCE). For approximately 200 years, the country was caught up in a pyramid-building frenzy. Pyramid construction required an immense amount of labor, skill, and organization. The Great Pyramid of Giza, for example, is made of 2,300,000 stone blocks—each weighing about 2.5 tons. Quarrying the stones, dragging them to the site of the pyramid, and laying them down would have taken 84,000 people employed 80 days a year (during the flood season, when farmers could leave their fields) 20 years to build.[26] Of course, even more people would have been necessary to supervise each stage of the construction, to engineer the structure so that it did not collapse, and to cook bread and brew beer to feed the armies of stonemasons.

Although the Egyptians invented writing about 600 years before the first pyramid was built, not a single text dating from the period of this frenzied construction describes how or why they were built. In Mesopotamia, kings used their wealth to build elaborate temples, enormous palaces, major irrigation projects, and to fund long-distance merchants, but they did not put much effort into their tombs (with a few spectacular exceptions). The Egyptians, by contrast, poured massive amounts of time and money into their burial places. It may not be a coincidence that the first writing in Egypt comes from labels in a predynastic tomb. Whereas the Mesopotamians were intent on using writing to administer life, the Egyptians used it to further their ideology of death. The pyramids probably symbolized royal power and the godlike nature of these early kings. The land of the dead was also the land of the gods, and it was there that earthly kings could join their divine brothers and sisters. Some Egyptologists have speculated that the pyramids represented the rays of the sun and that their construction helped link the pharaohs to the dominant sun god Re.[27] Whatever the explanation, there is no doubt that pyramid building served to further organize Egyptian bureaucrats and gave a definite boost to predictive sciences like mathematics.

ANDES: PATCHWORK POLITICS

In 1532, Francisco Pizarro and his troops marched into Cuzco, the capital of the Inca Empire, in search of their fortunes. Michael Moseley explains what they saw:

> By all accounts it was unbelievable—it was alien—and it was magnificent. The many distant buildings were clustered so close to the clouds that men and horses of the expeditionary force fought for breath in the oxygen-deficient altitude. Catching and reflecting the sun, towering stone walls shimmered with brilliant hues of gold and silver.[28]

The Inca Empire of Pizarro's day encompassed the Andes, the dry coastal desert of Peru, and part of the Amazon. It was probably larger than both the Ottoman Empire and the Ming Empire—the two dominant political powers in Europe and Asia at the time—and it was several times the size of a relatively small country like Spain, from whence Pizarro had come. Despite this, Pizarro managed to topple it using only 260 soldiers, with the help of gunpowder, trickery, and diseases like smallpox and measles, which the Spanish had already brought to South America, with devastating effect, long before Pizarro arrived.

No one, no archaeologist, historian, or even Pizarro himself, doubts that the Incan Empire was a state. Yet the investigation of the origins of cities and states in the Andes has proceeded slowly. Even now, after centuries of archaeological excavation, there is no consensus about when the first state evolved in ancient Peru. There are two reasons for this: first, Inca historians informed the Spanish that civilization had arrived only a century before the Spanish, with the Inca, and second, the experience of Peru does not conform to state development in "the old worlds" of Europe, Asia, or Africa. To disentangle when the state first developed in ancient Peru, and why, we must begin by asking why archaeologists do not ask this question today.

THE INDIVIDUALITY OF ANDEAN HISTORY: Following the Spanish conquest, Spanish officials recorded Incan histories from various Incan witnesses. Because these informers were not unbiased (as Incans, they naturally thought that their civilization was superior to all others), they emphasized the unique nature of the Inca state. They told the Spanish that before the Inca Empire conquered South America, there was no civilization. Additionally, because the Inca did not conceive of time as linear, but as circular, with certain events returning again and again, the Spanish misunderstood their accounts. These misunderstandings made their way into archaeology, where more confusion emerged. Quite simply, Andean history did not look like history elsewhere, certainly not like the history of Egypt, Mesopotamia, or Greece—or even Mexico. Although the Incan Empire was a state, it was one without recognizable writing, sometimes thought to be the primary characteristic of civilization (see Chapter 4). Also, like Egypt, it was a land with enormous monuments, but few cities. Some of these ceremonial complexes were even built without the benefit of agriculture.

ZONAL COMPLEMENTARITY: In our discussion of the rise of Uruk, we emphasized its position between many environments: the Persian Gulf, the irrigated banks of the Euphrates, the marshes, and the desert. States in ancient Peru arose in environments which were even more extreme. They flourished by using resources from all of these zones, a strategy that archaeologists term "zonal complementarity." Located close to the equator, the Andes, the second highest mountain range in the world, demarcated thousands of diverse microregions where different plants and animals could flourish. This makes it very different from other great civilizations:

> If thriving civilizations had matured atop the Himalayas while simultaneously accommodating a Sahara desert, a coastal fishery richer than the Bering Sea, and a jungle larger than the Congo, then Tihuantinsuyu [the Inca Empire] might seem less alien.[29]

The mountains, deserts, and tropical rain forests always maintained connections, and these connections helped fuel the development of states in different regions. We will begin by looking at a center that emphasized these connections and that was located midway between the coast and the mountains. Whether or not this ancient site, Chavín de Huantar (800–200 BCE), a city and a ceremonial center, was part of an ancient state is unclear, but the experiences there laid the foundations for future Peruvian states and reveal important aspects of Peruvian civilization.

ART AND RELIGION AT CHAVÍN DE HUANTAR: Chavín de Huantar, located between the coastal plains and the tropical forest, controls one of the only mountain passes between the Pacific coast and the highland Andean valleys. It was a regional religious center, perhaps the home of an oracle who was revered equally in the plains and the forest. At the heart of the settlement is a U-shaped temple complex, a popular form of coastal religious architecture. Surrounding this temple complex were neighborhoods of priests and temple administrators—perhaps 2,000–3,000 people lived here during Chavín's heyday (400–200 BCE), when the site covered 45 hectares. Although the architecture was built in the coastal style, the fantastic decorations of this temple are drawn from tropical plants and animals. Carvings show shamans, their mouths filled with jaguar teeth, snakes flowing from their bodies, and mucus dripping from their noses, celebrating rituals fueled by the hallucinogenic drugs which grow in the tropical forests. The Tello Obelisk, a 2.5 meter carved stone monument festooned with tropical images may illustrate a myth whereby the Cayman god gave tropical plants like peanuts, chili peppers, manioc, and achira to mankind. Caymans and tropical plants are foreign to Chavín, where potatoes are grown and alpacas are herded, but at home in the Amazon.

Art that reflects popular styles used at Chavín has been found for hundreds of kilometers along the Peruvian coast. Even the later religions of the highlands, including those of civilizations near distant Lake Titicaca, may derive from the Chavín cult. Unlike Uruk or Egypt, whose monumental buildings were funded by their abundant grain yields, Chavín's prosperity came from long-distance trade and the integration of agricultural systems in the plains, forests, and mountains into one system, comprising every-thing from llama-herding in the montane valleys to cotton plantations on coastal rivers.[30] The development of llama herding provided this complex world with a beast of burden, which could be used to carry such products as potatoes or cotton tunics between the different zones.

WAS CHAVÍN A STATE?   Chavín was not the capital of a precursor of the Inca empire; instead it was ritual center for a set of religious communi-ties; it resembled Rome during the Middle Ages (a religious center) more than Rome during antiquity (a political center). Like Uruk Mesopotamia or Egypt, Chavín had monumental architecture, produced by a society with multiple classes: goldsmiths, stone carvers, officials, priests, servants, and peasants. Although we do not have direct evidence for Chavín math-ematics, their well-constructed buildings show some familiarity with it, and their metallurgy reveals their knowledge of other sciences. A bureau-cracy would have been needed to maintain the elaborate irrigation struc-tures on the coast, as well as to administer the needs of both priests and pilgrims.

So was Chavín a state? Richard Burger, the excavator of Chavín de Huantar, believes that the Chavín civilization was as complex as any an-cient state, noting that Chavín "rivaled the classical Greek *polis* in size and beauty."[31] Adriana von Hagen, on the other hand, suggests that Chavín, although complex, never became quite urban, and although it was the precursor of later Andean states, it was never itself a state:

> Chavín de Huántar brought Andean civilization to the threshold of urbanism. It never became more than a ceremonial center, but it appar-ently sat at the top of a hierarchy of centers, clearly the leader and a place of major innovation.[32]

The example of Chavín shows that answering questions about the origins of cities and states around the world is a complicated exercise. Chavín has some attributes of a state; it was probably economically and socially stratified, but it may not have been politically stratified. In re-sponse to the questions raised by centers like Chavín and the cities of the Indus Valley, archaeologists now prefer to talk about "complex soci-eties," rather than cities and states. Our ideas about civilization, and about the origins of cities and states, come from our own experiences. We see Mesopotamia as the ultimate ancestor of the West. History be-gins at Sumer not just because writing was invented there, but because it begins for us there, because Mesopotamian cities, merchants, and litera-ture are recognizable precursors of Greek institutions and modern life.

Mesopotamian definitions, however, are not necessarily appropriate to complex societies on the other side of the world.

CONCLUSION

Just like agriculture, the state was invented independently on nearly every continent of the world, except Australia and Europe. All states share two characteristics; they are stratified both politically and economically. All emerged from societies with some ranking and some specialization—not from simple foraging or village societies. Historically, most states share many other features, including cities, writing, predictive sciences, centralized economies, government systems, and division of labor. How and why these characteristics developed within the first states and why they are common to all states are questions that archaeologists continue to pose, and which we will address in the next chapter.

Yet despite these similarities, early states were also quite diverse. Each state developed its own ideology and its own forms of expression. Despite widespread trade and possibly diplomacy, early Egyptian and Mesopotamian states differed in terms of their size, the level of urbanization, political structure, and religious ideology. This diversity is the rule, not the exception, for each type of society we have discussed: bands, villages, chiefdoms, and states.

Perhaps the most important trend since the end of the Ice Age in human society "has been the replacement of smaller, less complex units by larger, more complex ones."[33] Bands of tens of people have yielded to villages of hundreds, which, in turn, have yielded to chiefdoms of thousands and states of tens of thousands (and upwards). The invention of the state accelerated this tendency. The state is the most successful political form ever invented; during the last 6,000 years, states have managed to overwhelm and incorporate chiefdoms, village societies, pastoral nomads, and foraging bands. States have superior technology, larger armies, efficient command structures, and religious or patriotic ideologies that encourage fighters to sacrifice themselves for the good of the state. Indeed, the invention of the state represents an irrevocable development; once the state was invented, even if a specific state or empire collapsed, another would rise in its place.

TIMELINE

	6,000 BCE	5,000 BCE	4,000 BCE	3,000 BCE	2,000 BCE	1,000 BCE	0 BCE	1,000 CE	Present
West Asia	First irrigation (Choga Mami)	Ubaid chiefdoms: Eridu (Chiefdoms in Mesopotamia)	The world's first city: Uruk (States in Mesopotamia)						
East Asia									
South Asia									
Central Asia									
Southeast Asia									
Australia									
Oceania									The rise of the Hawaiian state
Europe									
North Africa		Egyptian chiefdoms (Nagada, This and Hierakonpolis)	Egyptian state						
West Africa									
East Africa									
South Africa									
North America									
Meso-america					Mesoamerican chiefdoms: Olmec and Zapotec (La Venta and San José Mogote)	Mesoamerican chiefdoms: Mesoamerican states: Monte Albán			
South America				Peruvian chiefdoms: El Paraiso		Andean civilization: Chavín de Huántar			

CHAPTER FOUR

THE ELABORATION OF CIVILIZATION

OUTLINE

Reading, Writing, and Arithmetic
Mesopotamia: One Fish, Two Fish, Red Fish, Blue Fish
Mesoamerica: Managing Time, Writing History
China: Divining the Future
Andes: Messengers and Knotted Rope

Bronze Age Economics
Mesopotamia: Division of Labor and the Management
Revolution
Mycenae: The Government Versus the Merchants
Teotihuacan: Artisans, Armies, and Traders

Government and the Law
Mesopotamia: Primitive Democracy or Primitive Dictatorship?
Indus Valley: Who Needs a State Anyway?

Conclusion
Timeline

GETTING STARTED ON CHAPTER FOUR: The rise of complex societies over the last 5,500 years has stimulated a number of intriguing developments: the invention of writing, the development of economic regulations, the emergence of systems of government, and the appearance of nonagricultural professions. Why did all of these developments appear at the same time as the emergence of the state? Did they all have a single cause? Why, and how, was writing invented in Mesopotamia, Mesoamerica, China, and the Andes? Did early states have fundamentally different economic systems? Why or why not? Does the example of the Indus Valley, a complex society without a state, mean that we should change our ways of thinking about the rise of states? How have our modern preconceptions led us to misunderstand early complex societies?

On the day before his son, Pepy, left home to attend school, Dua-Khety, a scribe working for the Egyptian Pharaoh, gave him some advice. Dua-Khety was worried that Pepy, when away from his father's care, might spend his time enjoying the delights of the court rather than studying. To convince him that doing his homework would help him in the long run, Dua-Khety explained the alternatives to being a government-employed scribe—an Egyptian white-collar worker—to Pepy. His descriptions of the other Egyptian professions, from the coppersmith, whose "fingers were like the claws of the crocodile" and who "[stank] more than fish roe," to the field hand, whose "fingers are swollen and stink to excess" and whose only payment for all his work is continual "sickness," to the fisherman, who is described as more miserable than any other professional because "he labors on the river, mingling with crocodiles," chronicle the hard work and suffering of most Egyptians. He concluded his discussion of the horrors of manual labor with some advice on how his son could rise above it: "Hence if you know writing, it will go better for you than those professions I've set before you, each more wretched than the other . . . the day in school will profit you, its works are forever."[1]

Egyptologists date this composition to about 1900 BCE. Dua-Khety, its supposed author, may never have existed—but his advice and worldview appealed to many Egyptians. No fewer than 40 copies of this piece have been found in Egypt; many of them date to 1500 BCE, 400 years after this tale was first written.

This story illustrates some of the changes that the urban revolution wrought and that will be explored in this chapter: writing, the invention of professions, and the creation of an educated elite. The development of the state did not simply produce ranks of bored bureaucrats or a rigid class system; its unintended consequences included the development of literacy, monumental architecture, division of labor, and varied political systems. These elements combined to form distinct complex societies, or civilizations. We will use Dua-Khety's perspective as an inhabitant of one early complex society to explore how some of these elements came into being in different places around the world. We will consider why writing was invented for different reasons in Mesopotamia, Mesoamerica, and China, and try to decide whether the Inca Empire had a writing system. We will also examine how early states structured their economies in response to different problems and opportunities. Finally, we will consider how and why different political and social systems emerged, including democracy, monarchy, and the caste system.

READING, WRITING, AND ARITHMETIC

Dua-Khety was not alone in seeing writing and schooling as the linchpin of civilization. Traditionally, western historians have seen the invention of writing as the boundary that separates prehistory from history. This distinction is echoed in China, where the Mandarin word for civilization, *"wen hua,* actually refers to the 'transformative power of writing.'"[2] In Mesopotamian legend, the tablet of destinies, a powerful document that the gods guarded, controlled the operation of the world, allowing wood to be made into weapons and people to live. Almost all early literate peoples, including the Mayans, Chinese, Mesopotamians, and Greeks, believed that writing was a gift from the gods, or from a semi-divine sage.

DEFINITION OF WRITING: True writing records a spoken language, including grammar and phonology (the sound elements of a language), not just meaning. A series of pictures showing a man, two rabbits, a fire, and an appetizing plate of food can be understood by speakers of any language Written language is different; although it can consist of the same symbols, it must also contain grammatical elements, which indicate, for instance, whether the man is cooking the rabbit now, or if he did so yesterday, as well as phonetic signs. Think of the difficulty of recording English solely in pictures, particularly when it comes to drawing prepositions or conjunctions. Some written languages began with *pictograms* that could also be read as sounds. An image of a bee, for example, could stand for both the insect as well as for the word *be.* Using a combination of *logograms,* signs that represented words, like god, as well as *phonograms,* signs that represented the syllable *an,* scribes could write anything from shopping lists to philosophic treatises. Other written languages used other symbols, like an alphabet or knotted rope, for example, which also recorded all elements of a language.

IMPACT OF WRITING: Writing extended the collective memory of a society. Before writing, events were lodged in the memory of individuals. After writing, those events could be recorded and consulted by anyone who could read—in theory, but seldom in practice, an unlimited number of villagers. It allowed information to be sent to the ends of the earth, making the spread of religions easier (it is not a coincidence that fast-growing missionary religions such as Christianity and Islam are based on books), and encouraging the elaboration of diplomacy. Early kings may have won their kingdoms through military conquest, but they ruled them by means of bureaucrats and a few newly invented tools: reading, writing, and arithmetic.

Not all complex societies invented writing for the same reason; instead, it was used initially for administrative, religious, and historical purposes. Archaeologists and linguists often state that writing emerged independently in only three places in the world: Mesopotamia, China, and Mesoamerica. All other writing occurred because of contact with one of these three centers. We will begin by exploring how and why writing developed in these early civilizations and conclude by looking at whether or not writing developed in the Andes.

MESOPOTAMIA: ONE FISH, TWO FISH, RED FISH, BLUE FISH

THE WORLD'S FIRST WRITING: The first true writing in Mesopotamia dates to about 3500 BCE and comes, rather fittingly, from the Eanna precinct of Uruk. At the time that the first writing was invented, the Eanna temple complex was the economic heart of a nascent state. Unimaginable wealth had gone into the construction of these lavish buildings, while more resources poured in constantly to maintain the gods (and the priests) of the temple in the style to which they had become accustomed. The world's earliest writing, tablets that contained lists of goods received by the temple, began as a way to keep track of all the items and people constantly going in and out of the temple. In Mesopotamia, the beginnings of writing served much the same use as today's Post-it notes—they began as a way of reminding other office workers what had been done: who had brought in the sheep for sacrifice, whether monthly rations had been paid out to the snake charmers and scullery maids.

We know how and why writing was invented in Mesopotamia because we have evidence of an ancient symbolic system that evolved into cuneiform. In most other areas of the world, our first evidence of writing postdates the actual invention of writing by an unknown period of time. This is because people usually wrote on perishable material—papyrus, bark, or wooden writing boards. In resource-poor Mesopotamia, however, the Sumerians, happily for archaeologists, chose to write on clay. The first tablets from the Eanna precinct use a fully developed writing system. Although we do not have tablets where the scribes sketched tentative, nonstandard signs, we do have archaeological evidence in the form of distinctive clay tokens and clay envelopes of the administrative system that chiefs and traders used before the invention of writing.

THE INVENTION OF THE TOKEN SYSTEM: Farmers in Mesopotamia began using clay tokens to keep track of agricultural and pastoral products soon after agriculture became widespread at around 8,000 BCE. These first tokens, which came in shapes like tiny spheres, cylinders, and cones, probably represented an early system of counting goods on a one-to-one basis. One cone, for example, meant one basket of grain; one cylinder meant one sheep. Abstraction in counting—envisioning the concept of "oneness" or "twoness"—came later with the origins of writing.

FROM TOKENS TO TABLETS: As social and economic systems became more complex, with the development of Ubaid chiefdoms (5500–4000 BCE) and Uruk states (4000–3000 BCE) (Chapter 3), the token system also increased in complexity. Some new token forms appeared, including animal heads and miniature jars, while many of the old geometric tokens were now decorated with lines scratched into their surface. The invention of complex tokens coincided with the appearance of new craft goods. As Denise Schmandt-Besserat, the world expert on the token system explains, they kept track "of products for which Mesopotamia was famous: textiles and garments; luxury goods, such as perfume, metal, and jewelry; manufactured goods, such as bread, oil, or trussed ducks."[3] A new system of managing tokens also appeared. Managers placed simple tokens in clay envelopes, which served as archives, preserving the information for future use. In order to know the contents of the envelope at a glance, Mesopotamian entrepreneurs marked the envelope's surface, either by pressing the tokens into them or scratching and notching the clay. At some point, these marks became more important than the counters themselves, and finally supplanted them. Token-filled envelopes disappeared, while tablets with token impressions or scratch marks became common around 3500 BCE. The new tablets communicated the same information as the tokens. We have the token predecessors to many of the first signs: characters like food, oil, honey, beer, cloth, and grain.

THE INVENTION OF LITERATURE AND MATHEMATICS: Using the tablets led to transformations in Sumerian mathematics and literature. Four hun-dred years after the first tablets appeared, at approximately 3100 BCE, there were tablets that exhibit abstract numbers (numerals like our own 1, 2, 3, 57, 1206, etc.). These numbers are impressed, just the way the simple tokens were, but rather than representing one jar of wheat, one field, or the like, they represent cardinal numbers and are followed by scratched pictographs that represent the commodity. Once the picto-graphs no longer represented numbers as well as products, they could communicate a dizzying array of subjects—not just goods managed by an accountant. The invention of abstract numbers may have encouraged some scribes to abstract other nouns as well and invent symbols that stood for concepts like "mouth," or "speech," or "to talk." The desire to record the names of the people who used this system, like the pious shepherd who donated lambs to the temple (or the chef who roasted them), led to the invention of a syllabary, where symbols represented syllables as well as objects. Once this happened, writing no longer served just one purpose, that of recording economic data; now it could be used to compose hymns or letters or record any form of speech. The first examples of literature, a series of hymns to a temple, and the first his-torical documents, royal inscriptions, appear at approximately 2600 BCE, long after the invention of writing.

In Mesopotamia, writing emerged from accounting, and the elaboration of literature occurred at the same time as the elaboration of mathematics. The invention of numerals allowed officials to solve simple equations. City planners used writing and abstract numbers to calculate how many bricks were needed for a city wall, or how much barley seed was needed for a field of a certain length and width. Priests soon realized that by observing the movements of the stars they could devise a more sensitive calendar than simply using the phases of the moon. By recording this information, and building on that recorded knowledge, they invented a science that in later Mesopotamia (Babylon) reached great heights and became the foundation of our modern discipline of astronomy.

Tokens, however, did not entirely disappear with the invention of writing. Ordinary people, who were never taught the new system, continued to use them. Shepherds kept tokens recording the number of sheep in their care, while farmers used tokens to chronicle yields from previous years. The latest set of tokens ever recovered dates to 1600 BCE, nearly 2,000 years after the invention of writing.

MESOAMERICA: MANAGING TIME, WRITING HISTORY

The earliest writing in Mesoamerica emerged from a world of complex chiefdoms on the threshold of state formation. Although we do not have the predecessors for writing in Mesoamerica, recent finds in Mexico show us two episodes in the invention of writing: an enigmatic early stage between 1200–900 BCE when writing emerged in Olmec territory and then disappeared without a trace, and a better-documented second stage from 750–250 BCE, when writing appeared nearly simultaneously in both the Zapotec and the Olmec territories. The writing from this later stage has clear links to later Mesoamerican writing; early examples of both of these later scripts record dates from a pan-Mesoamerican, sacred calendar. Unlike in Mesopotamia, where writing was first and foremost an administrative device, whose origins were tied up with accounting, in Mesoamerica scribes wrote in order to manage time and record historical events.

OLMEC WRITING, 1200–900 BCE: In 2006, the Mexican government asked a group of archaeologists to examine an inscribed stone block from the Cascajal quarry in Veracruz, Mexico, not far from the Olmec center of San Lorenzo. When the archaeologists visited Cascajal, they were surprised to learn that the objects found with this block all dated to between 1200 and 900 BCE, making this the oldest writing ever found in Mesoamerica by nearly 500 years. Unfortunately, since the block was not found in an excavation, archaeologists cannot be certain of its date, nor can they hazard many guesses about what the writing may say. The writing, which consists of 62 signs, probably conveys a complex message, such as an inscription celebrating an Olmec king's achievements. Mysteriously, the writing that covers this stone does not resemble later Mesoamerican scripts, implying that it flourished and disappeared before writing appeared across Mesoamerica in the middle of the first millennium BCE.[1] Later Mesoamericans seem to have credited the Olmecs with the invention of writing. The words for "to write," "paper," "year," "to count," and "twenty" (the basis of both the counting and calendar system) used throughout Mesoamerica were probably borrowed from the language spoken by the early Olmecs, attesting, like the newly discovered block, to the antiquity of Olmec writing.

OLMEC WRITING, 750 BCE: Our earliest evidence for the second stage of writing in Mesoamerica also comes from Olmec territory, although in this case the writing has clearer links to later Mesoamerican scripts. In 2002, archaeologists excavating the site of San Andrés, a regional Olmec center only five kilometers north of La Venta, Tabasco, found a cylinder seal and fragments of a serpentine (green stone) plaque, both featuring hieroglyphics, among the debris from a major feast in a level dating to 750 BCE. A bird with a speech bubble issuing from his beak containing glyphs (units of writing), which read "King 3 Ajaw," was carved into the cylinder seal. 3 Ajaw is probably both a date in the Olmec calendar system, and following Mesoamerican tradition, the name of an Olmec king. Most Mesoamerican civilizations used two different calendars: a special 52-year "calendar round" which was the result of meshing together the 365-day solar year and the 260-day ritual calendar, and the "Long Count," a calendar which counts the days which have elapsed since the starting date 13 August 3114 BCE (much the way our own Gregorian calendar begins on January 1, year 1 CE). The inscriptions from Cascajal and San Andrés thus suggest that the Olmecs invented two of the most important Mesoamerican cultural features, the calendar system and writing.

ZAPOTEC WRITING, 500 BCE: Far from the Mexican coast, in the Oaxaca valley, the Zapotecs also developed a system of writing. The earliest Zapotec writing dates to 500 BCE and consists of a date in the Mesoamerican calendar system, "1 Earthquake," carved into a stela depicting a prisoner of war at San José Mogoté (Figure 4). Like in the Olmec example, 1 Earthquake probably represented both the name and birth date of this prisoner of war. His portrait is the earliest example of Zapotec writing, but by no means the only. Other *Danzantes*—the name given to carvings of mutilated corpses on stone slabs—also contain dates, probably the names of the victims of Monte Albán's superior warriors. Two other monuments display an entire column of writing without any corresponding figures, suggesting that like early Olmec writing, Zapotec writing could also be used to convey more complex messages. The Zapotec language remains to be fully deciphered, so we cannot read these messages yet. In the following centuries, from 250 BCE–700 CE, when Monte Albán emerged as a prominent Mesoamerican center, Zapotec writing adorned stelae recording both military and diplomatic victories.

WRITING AND KINGSHIP: For both the Olmecs and the Zapotecs, it seems that writing and the calendar system were tied up with the institution of kingship. Olmec cylinder seals were used to imprint designs on clothes or the body for decoration. According to Mary Pohl, the excavator of San Andrés, "imprinted cloth or body decoration and engraved jewelry such as the plaque were [some] of the principal means by which high-status individuals conveyed the message of kingship."[4] The Zapotec *Danzantes*, many of which adorned a temple in Monte Albán, also exhibited the high status and power of early Mesoamerican kings through both images and writing. Javier Serrano, a Zapotec hieroglyphic specialist, writes that "in Mesoamerica, as in many other parts of the world, writing was imbued with divine origin and messages."[5]

CODICES, ECONOMIC DOCUMENTS, AND LITERATURE: But is this the whole story? We are almost certainly missing the vast majority of Mesoamerican records.[6] The earliest writing in Mesoamerica contains elaborate calendar glyphs, war accounts, and diplomatic ties, not because writing was used only ceremonially, but because ceremonial writing is all that survives. Mayan artwork shows later Mesoamerican scribes writing on codices—books made of bark. We have only three Mayan codices because they were made of materials that disintegrated over time. Yet we know that codices were widespread in Mesoamerica because soon after the conquest, the Spanish burned thousands of Aztec codices, believing that they were the devil's works. Mayan and Aztec codices did not just record ceremonial information, but may have contained tribute lists, astronomical information, and works of literature, much like the tens of thousands of Mesopotamian accounting tablets (which were not, unlike the stone Danzantes, made to last, but which survived due to unusual conditions). Although this evidence dates to a

FIGURE 6
Oracle bone from Shang times with an inscribed question and cracks caused by exposure of the bone to heat. Shang Oracle Bone.

period much later than our first evidence for writing, scribes probably recorded economic information on utilitarian, easily perishable materials from the beginning. The Mayans had a "certified public accountant" god from the early Classic period (ca.150–450 CE), an indication that writing was not far from accounting in this society, as well as in Mesopotamia.

CHINA: DIVINING THE FUTURE

THE EARLIEST CHINESE WRITING: Unlike Mesopotamian and Mesoamerican writing systems, which were rediscovered and deciphered by scholars during the nineteenth and twentieth centuries, the Chinese writing system remains in use. More than a billion people read and write Chinese, using characters that are the clear descendants of pottery marks that date back to the third millennium BCE. The earliest recovered traces of writing in China are on pottery sherds, *oracle bones* (ox shoulder bones and turtle shells), and bronze vessels dating to around 1500 BCE, but, as in Mesoamerica, there are indications that scribes usually wrote on perishable material—in this case, bamboo strips. However, the oldest preserved wooden strips with Chinese characters come from the fifth century BCE, 1,000 years after our inscribed oracle bones. Like the bark codices on which the Zapotecs, Olmecs, and Mayans wrote, most of these have long since decayed.

LINEAGES IN EARLY CHINA: Writing was invented in China to communicate with the ancestors, for ritual purposes in a clan-oriented society. Although we do not and will never have (barring future archaeological finds from extraordinary contexts) all the evidence for the story of the origins of writing in China, the pieces we do have reflect the place of writing in early China. In the first Chinese states, the basic building block was not the individual, but the family. Each person belonged to a lineage, which in turn belonged to a clan. Lineages and clans consisted of both living members and dead ancestors; the latter were propitiated in a number of rituals. Sometime during the chiefdom phase in China (the Longshan period, 2600–1900 BCE), the most powerful lineages of each clan stopped marrying and interacting with other members of their clan and started marrying each other instead, providing the first elite. During the Shang Dynasty, the period from which we have the first writing, the most important lineage of the most important clan produced the kings, who claimed that they were descended from the high god Shang Di. The eldest son of the king inherited the throne and ruled the main city, while his younger brothers either received lower positions at court or new territories where they built their own capitals. The eldest sons of these younger brothers would inherit their thrones, while the younger brothers could branch off and found new towns. In this way, the land that the early Chinese state ruled continually expanded. In less important lineages, professions were passed down from father to son, allowing lineages to coincide with professional groups like guilds. This basic pattern seems to have been in place during the first three dynasties: the Xia, the Shang, and the Zhou.

CLAN SIGNS AND VESSELS: As may be expected in a society where clan membership was all important, the first traces of writing have been interpreted as clan signs. These "clan signs" are simple characters found on pottery vessels from sites in the Yellow River valley dating from 6000–2000 BCE. They are not true writing, because they do not necessarily record any language, but are instead "potter's marks" used to distinguish the person (or the lineage of the person) who made the vessel. By the time of the Dawenkou culture (4000–2500 BCE), on China's east coast, many large pottery jars had several signs inscribed on them, which archaeologists believe are clan identification marks. Even during the Shang period, when scribes kept the majority of their records on other materials, they continued to write short inscriptions on bronze vessels. By this time, we can read the inscriptions, which are often dedications to a particular ancestor of a particular clan, whose profession may also be noted. Occasionally they mention clan members' battle victories or participation in religious festivals. These bronze vessels once contained food or alcohol, which pious Shang men and women offered to their ancestors during feasts both to secure their favor and to brag about how well their descendants were doing. It seems likely that the earlier pottery vessels were made for a similar reason. In the Neolithic period, gifts of food were an important way to cement social alliances between living relatives and the dead. People began writing on these food vessels to communicate with their ancestors more directly.[7] It is fitting, then, that the first symbols identifiable as a full writing system are long inscriptions on pottery sherds, which have not yet been deciphered. Nevertheless, we can read one character, found on a wine jar fragment dated to 2000 BCE, as *wen*—which means both culture and writing in Chinese.

SHANG ORACLE BONES AND ANCESTOR WORSHIP (1766–1027 BCE):   The first understandable texts written in Chinese are on oracle bones, so called because priests and diviners interpreted the pattern of cracks caused by exposing these bones to fire as predictions from the ancestors. These oracle bones were carefully archived at Anyang, the last capital of the Shang dynasty. In some ways, these oracle bones bridged the gap between the purely ceremonial records of the early Mesoamerican civilizations and the purely functional records of Mesopotamia by having both archival and ritual functions. Like the pottery inscriptions, the oracle bones are an attempt to communicate with the ancestors, to get reliable predictions about the most important and always uncertain events in any state: child-birth, war, and harvests. Priests carried out the divination rites with oracle bones during feasts, which served to cement the power of the king. Trib-ute from dependent states supplied the turtle shells used for these rites, while the oxen which supplied the scapulae came both from the king's herds and from distant lands. Unlike writing in the Near East, which was always concerned with accounting, writing in China seems to have its roots in kinship organization. As Anne Underhill explains, "this may give us a strong hint of ancient Chinese priorities: membership in one's kin group was the first thing that the first writing recorded, because it was key to the ancient Chinese social order."[8]

ANDES: MESSENGERS AND KNOTTED ROPE

When Pizarro conquered the Andes, he defeated an empire that was heir to more than 2000 years of civilization, from Chavín de Huantar to Tihuantinsuyu (Chapter 3). In fact, complex societies may have flourished longer in the Andes than in Spain. Nevertheless, in discussions of ancient states and empires, Andean civilization usually remains a strange foot-note, the only complex society that never invented writing. Instead of scribes, the Incan Empire trained specialists to encode and decipher a sys-tem of knotted strings that they called *quipu*. Quipus recorded how much tribute each dependent town and province could supply to the great capi-tal at Cuzco, as well as other numerical data. The quipu system worked so well that the Spanish continued to use it to gather taxes and tribute for 50 years after they conquered the kingdom of Peru. In fact, it was only in the 1580s, when the Inquisition decided that quipus were "objects of idol-atry" that must be destroyed, that writing (in the form of Spanish records) truly replaced quipu in the Andes. How did quipus work? Is this system really distinct from writing?

DECIPHERMENT OF NUMERIC QUIPU: The decipherment of the quipus began in 1912, more than 300 years after the last quipu was constructed, when L. Leland Locke, an American anthropologist, used an account of the Inca number system, written in Spanish in the sixteenth century by a native Incan after the Spanish conquest, to analyze a quipu preserved in a museum. He deciphered the information in the quipu and showed that the Inca could use it both to solve simple arithmetic problems and to analyze geometrical concepts, like calculating the area of a circle. Most quipus were ancient balance sheets containing the income and expenditure for a city during a particular period of time. The top cord of a quipu carried the grand total, whereas other strings allowed the quipu readers to see the sum of certain rows or columns at a glance. As in Mesopotamia, the Inca recording system began as a system of accounting. Unlike in Mesopotamia, however, many Andean specialists would propose that it ended there as well, and that the Inca never went on to develop true writing.

EVIDENCE FOR QUIPU AS WRITING: There are a few problems with this easy dismissal of the quipu as a bookkeeping device. First, several Spanish accounts of the Inca Empire indicate quipus were not just used as a reckoning system, but also recorded history. Most historians have assumed that the placement of the knots on the quipu reminded the storyteller of a story that he had memorized, but did not actually contain the account; they served as a memory aid for a storyteller schooled in a rich oral tradition. A new generation of quipu scholars disagrees. They point to Spanish accounts from the early days of the Spanish conquest (1532–1585 CE) that the authors insisted were Spanish translations of quipu readings. These translations, stored in colonial records offices in Spain, may be the clue to deciphering the remaining 600 quipus held in museums around the world (and the others that occasionally surface in archaeological excavations). The standardization of the quipu translations has led historians to believe that the different combinations of knots and strings actually represented words or syllabic elements, and that the quipu could be read in the Inca language, Quechua. This makes the quipu system a form of writing (albeit an unusual one). The decipherment of the quipu system, if it does record language, would give archaeologists and historians a unique insight into a now dead civilization.

If quipu is a form of writing, then why has it taken scholars so long to recognize it? The answer lies partly in the same Spanish accounts that we now take as evidence that the Inca did have writing. In the 50 years following their conquest of the Inca Empire, the Spanish were engaged in a war of propaganda, desperately trying to persuade their new subjects of the superiority of Spanish over Inca rule. One way of doing this was to discredit remaining Inca practices—such as the quipu. At the same time that they used quipu to record the collection of tribute and as evidence in courts of law, the Spanish derided the quipu readers as liars. As Gary Urton explains, "In the strife-ridden, highly antagonistic world of the colonial

Andes, characterized from the beginning by the domination of natives, the Spaniards could not allow the authority of native records and record-keepers to stand for long."[9] Because the Spanish were so definitive about the limited use of the quipus, and their inferiority to written Spanish records, archaeologists treated these objects the same way—as mute artifacts, not records. It is only now that we are beginning to address this misbalance.

BRONZE AGE ECONOMICS

We can understand Dua-Khety's bitter denunciation of manual labor because the professions we choose define our lives no less than his. Before the invention of complex society, however, this was not true. Among hunter-gatherers and simple villages, everyone shares the same basic knowledge and engages in the same basic activities. Sometimes labor is divided by age or gender, but these definitions are fluid, changing from one group of foragers or villagers to the next (Chapters 1 and 2). Chiefdoms may allow a small amount of specialization by paying skilled craftsmen a "salary," but even these artisans make pots or knap flint only part time and spend the rest of their time hoeing their gardens or fattening their pigs (Chapter 3).

With the development of complex societies, this all changes. States demand large amounts of manufactured goods—to decorate their palaces and display their importance—and are willing to fund their production. The invention of professions, however, encompassed more than just allowing certain people to work full time as blacksmiths or jewelers or potters. The new division of labor included the development of a professional military, composed of trained soldiers equipped to defend the state and conquer new territories. It also promoted the invention of "management"—a profession whose practitioners serve the state's interests, but manufacture nothing.

We must be careful, however, not to overstate the case. The Egyptian satire of trades, the Sumerian standard profession lists, and Indus valley seals all illustrate the wide number of careers open to people from 5000–3500 years ago—but these early states were still far from our own hyperspecialized world. Most families still lived on their farms and were close to self-sufficient. Even people who worked as soldiers or builders for the state might do so only for a few months out of the year. Although these early urban economies incorporated many novel ideas, they also relied upon earlier traditions. To see how and why early career life began in the first cities, we will consider the interplay between the creation of a division of labor and the emergence of the state in Mesopotamia, Mesoamerica, and Mycenae.

MESOPOTAMIA: DIVISION OF LABOR AND THE MANAGEMENT REVOLUTION

Trying to determine how people made a living in fourth millennium BCE Mesopotamia is a tricky matter. Written documents that record the payment of rations to a wide range of artisans have encouraged historians to talk about the origins of professions. Seals show rows of pigtailed women working in assembly-line fashion in textile factories or dairies. Others depict men carrying tools to a building site, strong evidence of an economy where not everyone farms. At the same time, the archaeological evidence indicates that a lot of communities in southern Mesopotamia were still self-supporting. They made their own pottery and stone-tools, and grew their own food as well. How did the urban revolution stimulate the rise of a specialized economy? How did the inventions of professions change the lives of ordinary people?

THE RISE OF URBAN WORKERS: Different communities always offer different opportunities to people. Certain communities had probably specialized in some products from early periods; their interaction helped produce the Mesopotamian state (Chapter 3). What transformed the situation were the actions of temples like those found first at Eridu, and later at Uruk. As the temples began demanding tribute—both in goods (sacks of grain for the gods) and in labor (workers to till the gods' fields)—two things happened. First, the temple was able to use what it got in tribute to pay people to do something besides grow grain, raise sheep, or fish for carp. Second, the tribute demands proved to be too much for some farmers. They had a hard enough time feeding their families, let alone supplying the gods with 10 percent of everything they grew. Some people left their lands and fled to the growing cities—even to the temple itself—to escape from tribute obligations. These people provided the workforce for new large-scale endeavors. At the same time, the temple probably attempted to attract (or simply adopted) individuals who showed talent for different crafts.

INCREASED EFFICIENCY AND LABOR-INTENSIVE PRODUCTION: Urbanism promoted efficient production in some industries and labor-intensive production in others. Much like their twenty-first century counterparts, Mesopotamian executives promoted efficiency by encouraging innovative technology (such as the potter's wheel or drills for mass-producing cylinder seals) and by using labor more effectively. In potters' workshops, for example, the work was standardized. Fingerprint analysis shows that many people were involved in the production of one pot. One potter prepared the clay, another threw the vessel, and a third carried the vessel somewhere to dry. Groups of potters began using huge, state-run kiln installations, rather than small household kilns. The temples also encouraged labor-intensive activities, such as the manufacture of luxury goods and monumental buildings. Why did these two opposite trends appear simultaneously? The new landless laborers pouring into the nascent cities had to be kept busy, so they did not overthrow the system that had stolen their land. As one archaeologist explains, "the need to create projects to employ temporary surpluses of labor may have resulted in the deliberate attempt to 'make work,' leading to the construction of ever larger and more elaborate monuments and the production of increasingly fancy luxury goods."[10]

WOOL AND THE LOSS OF WOMEN'S ECONOMIC POWER: Once started, the professional revolution snowballed, and newly created industries employed increasing numbers of people. These industries caused changes to daily life, particularly the lives of women, which had far-reaching consequences. In the fourth millennium, selective breeding of wooly sheep allowed people to use wool on a large scale for the first time. Prior to this, most people's clothes were made out of linen, a fabric that comes from flax plants. Early farmers domesticated flax by 9000 BCE, and for more than 5,000 years, individual women grew it in their gardens and made clothes for their families. It takes a lot of work to make linen out of flax; women must water, weed, and harvest the plants, extract and beat the fibers, and then spin and weave the linen. All in all, it takes roughly 57 days of work to produce enough linen to make clothes for one person for a year. It is much easier, on the other hand, to turn fleece into wool. The key difference is the way that sheep are raised. Three shepherds can easily manage a flock of 100 sheep, which can graze in the steppe (unlike flax, which must be grown on the best agricultural land). Those 100 sheep can produce enough wool to clothe 40 people per year. When women grew their own flax, they controlled the entire process of making cloth. As soon as male shepherds started supplying the wool, women lost access to raw materials. State officials began serving as middlemen, taking the wool plucked from the sheep and distributing it to weavers. In Mesopotamia, the loss of control over flax, combined with heavy tax demands from temples, forced poor families to sell women into debt slavery. Once in the cities, they worked in textile workshops that employed thousands. Weaving and sewing became big businesses, and women lost control of their labor.

THE RISE OF THE HOUSEHOLD SYSTEM: The development of a labor economy is at the heart of the transition to the state. Scholars like to envision the state as an organization that severed family ties; tasks that had been previously performed at home were taken out of the control of individual families and became state industries. The poor women pictured on the Uruk seals, who spent their days in a workshop, carding and weaving wool far from their families, lived in a way unimaginable to women before the invention of the state. Yet the state continued to use the same kinship and household terminology to describe itself, as people in earlier societies had done. The Mesopotamians termed all economic units, including those run by temples, which might employ thousands of people, "households." The heads of these vast corporations called themselves "fathers" and their employees "children." As a result, someone like the queen of Lagash, who ran a household that employed thousands of people—artisans, farmers, servants, and bureaucrats—thought of her concern as not qualitatively different from a small peasant household in one of the nearby villages. In her mind, the difference was probably one of degree. Both households were "self-sufficient" and relied on the division of labor. For the peasant household, this meant that the mother raised the children, hoed the garden crops, fed the pigs, and cared for the flax, and the father ploughed the fields, sowed the barley, and repaired the house. For the queen of Lagash, however, this meant that thousands of people worked at specialized tasks. This model of the household lasted for a long time; the classical Greeks considered the basic economic unit to be the household (*oikos*), a word which is the root of our own term "economy."

MYCENAE: THE GOVERNMENT VERSUS THE MERCHANTS

THE DECIPHERMENT OF LINEAR B: In 1952, a young architect with a linguistic bent deciphered the writing on hundreds of mysterious clay tablets that had been discovered among the ruins of palaces on Crete and the Greek mainland. These palaces, built of massive stone blocks (called Cyclopean, because later Greeks thought they had been made by one-eyed giants), were the remains of a civilization that flourished and disappeared almost 1,000 years before the better-known city-states of classical Greece. We call this civilization *Mycenaean*, after Mycenae, the most important city on the Greek mainland from 1600–1150 BCE. Before 1952, scholarly reconstructions of Mycenaean society and economy drew on Homer's *Illiad* and *Odyssey* and were filled with wild speculations. The decipherment of Linear B (the Mycenaean script) allowed fantasy to yield to reality by revealing the records of the ancient Mycenaeans. To scholars obsessed with poetry and treasure, the Linear B tablets were disappointing. They contained no literature, philosophy, or history. Instead, the Linear B texts were careful records made by the administrators and accountants charged with overseeing the palace's finances: tax lists, production quotas, inventories, and payrolls. In the normal course of things, these tablets were written, archived, and then discarded. The Linear B tablets that we have were preserved because they were in use when the Mycenaean city-states were destroyed and Mycenaean institutions collapsed. They are like the dusty files of a Soviet bureaucrat from 1991; not exciting at first, but they may illustrate both how this state functioned and why it collapsed. These tablets show us how the palace economy, which employed artisans, administrators, farmers and servants, worked. Archaeological evidence from Mycenae, however, hints that much of the economic activity in this society occurred outside of the state's grasp.

THE PALACE ECONOMY: All the Linear B texts so far discovered have been found in palaces, which tells us something important about Mycenaean society and the economy. In general, the capital cities of Mycenaean Greece were very small, one-fifth to one-twentieth the size of cities in contemporary Mesopotamia. They were fortified, and completely dominated by the palace, which was the economic heart of the Mycenaean state. The palace employed ivory carvers, mural painters, bronze workers, chefs, courtyard sweepers, and administrators. Outside of the cities lay several villages and farmsteads, with few dependent towns. This meant that the palaces within the cities interacted directly with the farmers and shepherds outside of them. Palace officials oversaw slaves who grew wheat and figs on palace land; they kept track of the shepherds who cared for thousands of palace sheep.

Of course, not everyone was part of this palatial system. Archaeological remains show that outside of palace-owned fields, farmers grew a diverse array of crops, just as their ancestors had, including barley, peas, lentils, and figs. Similarly, although the palace only records male sheep (kept for wool), villagers also kept female sheep (for breeding and milk). Nevertheless, the palatial net of control spread wide; those villagers who did not till its fields and were not fed from its stores had to supply the palace with a number of rare commodities and some services through taxation. In this way, the palace always had enough flax for linen, bronze for weapons or tools, hides, honey, and spices, as well as soldiers. No one "paid" for anything within this system; the concept of money was completely foreign to the Mycenaean consciousness. There are no words on the tablets for the economic activities that we practice daily in modern market economies, like "buy," "sell," "lend," or "pay a wage."[11] Similarly, despite the importance of the market place in Classical Greece, there is no evidence, either written or archaeological, that it existed in the Mycenaean world. People still traded different commodities, but this activity was not ritualized.

LONG-DISTANCE TRADE: Although the tablets seem to record economic data in exhaustive detail (listing who cared for each of the 80,000 sheep of the Knossos herd, for instance), they remain silent about several important aspects of the Mycenaean economy, especially the Mycenaean role in long-distance trade. Large, colorful Mycenaean jugs, often decorated with scenes of marine life, have been found across the Mediterranean: in Greek islands, the Levant, southern Anatolia, and even Egypt. These large containers must have carried something that was valued in all of the Mediterranean civilizations of the time, perhaps extra-virgin olive oil or fine wine. Sometime around 1350 BCE, a heavy ship loaded with trade goods sank off the coast of Turkey. The ship was carrying cargo from a wide geographic area, from northern Europe to the western Mediterranean to sub-Saharan Africa, including 350 copper ingots, a ton of resin (used for incense), Baltic amber, tortoise shells, elephant tusks, hippopotamus teeth, ostrich eggs, jars of olives, and stacked Mycenaean pottery.[12] Mycenaean sailors may have been the middlemen in this vast trade network, as goods from Europe flowed to Greece and were then exported to Egypt and the Levant, where sailors picked up valuable commodities from these lands.

EVIDENCE FOR ENTREPRENEURS: But who controlled this trade? Were these Mycenaean government employees or private traders? There are no written documents that record how the gold and ivory that artisans fashioned in palace workshops in Mycenae or Knossos came to be there. Some archaeologists believe that the palace administered this trade, as they did so many other aspects of Mycenaean life. They argue that only the palace was rich enough and powerful enough to finance these ships. It is only after the collapse of the Mycenaean state that we find ivory objects, Near Eastern cylinder seals, and semiprecious beads in places other than Mycenaean palaces, or graves of Mycenaean kings, suggesting palace control of trade prior to their collapse.[13] Yet other archaeologists believe that independent merchants financed the ships and reaped the profits from this trade, since centralized states rarely engage in international trade, which is hard to control and risky. Some even believe that this trade, like later Classical trade (or even modern international trade), served as the foundations of a primitive market economy, and that the ship owners were hardy entrepreneurs. They argue that this international trade proves that the palace's reach was limited, and that even if words for buy and sell aren't found in the Linear B tablets, the Mycenaeans bought and sold goods all the time. If this trade was in the hands of private entrepreneurs who conducted it according to market principles, we must modify our views of the Mycenaean economy as tradition bound and state controlled.

In most early states, the main unit of government, usually either a temple or palace, encouraged the division of labor by providing craftworkers and bureaucrats with rations that freed them from needing to farm for themselves. After the state was invented, however, the institutions that allowed for division of labor became more widespread. At Mycenae, a developed division of labor probably allowed certain independent merchants to trade with people across the Mediterranean to secure food in exchange for their goods, without the help of the state (or their own farms). This was the beginning of a market economy, although anything like our own modern capitalist system was still far in the future. Market forces may have set the prices for the luxury goods that Mycenaean traders plied in the eastern Mediterranean. On the other hand, the palace remained the most important economic institution—the biggest employer and the biggest consumer. It would not do, however, to overestimate the importance of either the state-controlled economy or the developing market economy. Outside of the confines of the palace and Mycenaean cities, people's agricultural activities followed traditional patterns, deviating little from pre-state practices.

TEOTIHUACAN: ARTISANS, ARMIES, AND TRADERS

The sheer size of the ruins of Teotihuacan, located north of Mexico City, still amaze the tourists who flock there yearly; the city spreads out over 22 square kilometers. Around 500 CE, it housed 120,000 people, many of whom lived in at least 2,200 apartment complexes. By contrast, Rome, which admittedly was experiencing a low point in its 1,000 year history, only covered 13.7 square kilometers and contained somewhere between 50,000 and 120,000 people. Similarly, Ctesiphon, the capital of an empire that encompassed Iran, central Asia, and even parts of the Indian subcontinent, had only between 23,000 and 49,000 people. Only in China were there two cities (Luoyang and Nanjing) that had larger populations than Teotihuacan. Teotihuacan was not only the most important center in Mesoamerica at that time—possessing more temples, setting the standards in fashionable living, and controlling the economy of a vast region—it was also one of the largest cities in the world. We can still sense this today when confronted with its ruins. The Pyramid of the Sun, which is located on the city's central avenue, towered above Teotihuacan. Although it is the largest structure ever built in the ancient Americas, it is only one of a number of monumental temples that dominated the heart of ancient Teotihuacan, a monumental district far larger than the entire city of Mycenae.

The inhabitants of Teotihuacan, unlike their Mayan or Oaxacan neighbors to the south, did not use writing. There are no economic tablets from the Pyramid of the Sun that mirror those found in the Eanna complex at Uruk 3,500 years earlier, nor are there stelae featuring the deeds of Teotihuacano kings as there are in many of the Mayan capitals. As a result, deciphering Teotihuacan economic history is a matter for careful archaeological analysis. As in Mesopotamia or Mycenae, Teotihuacan society exhibited a strong division of labor. The government probably employed thousands of soldiers and artisans, while traders and farmers flourished outside of the direct control of the state, selling their wares at the city's large markets.

MARKETS AND TRADERS IN TEOTIHUACAN: At the heart of this city, at the intersection of its four quarters, lay "the Great Compound," an open plaza that probably served as a marketplace—not a temple or palace as might be expected. Much of the archaeological evidence for Teotihuacan concerns ancient craft production and trade. The apartment complexes, for example, did not house just anyone, but were built for the city's craftspeople, who may have made up as much as 25 percent of the population. Foreign traders from other areas of Mesoamerica lived in distinctive neighborhoods. Traders and diplomats from Oaxaca, for example, directly imported some pottery from their hometown, while local potters produced other vessels in the Oaxacan style. They buried their dead in Oaxacan graves and carved (or imported) Oaxacan stelae inscribed with Zapotec hieroglyphics. Similarly, traders from Veracruz built their Teotihuacan houses out of adobe, in the same style popular back home. It is easy to imagine these traders plying their wares in the Great Compound, hawking luxury goods to the affluent, cosmopolitan residents of Teotihuacan.

THE OBSIDIAN TRADE AND THE RISE OF TEOTIHUACAN: Yet, the Teotihuacanos did not just import goods, they also exported both goods and ideas. Stone monuments in Tikal, a Mayan city in present-day Guatemala, show symbols from Teotihuacan, and a stone monument in Monte Albán's Main Plaza shows a peaceful meeting between this city's ruler and Teotihuacan's. A wide range of Teotihuacano manufactured goods, including pottery and obsidian blades, found their way to far-flung corners of Mesoamerica. Although agriculture and food distribution necessarily formed the largest part of Teotihuacan's economy, obsidian, which could be easily worked and traded long distances at low cost, fueled an expansive long-distance trade. Because obsidian weighs little, takes up little space, and is worth a lot, individual traders or small caravans can easily carry it. Exporting ears of corn, on the other hand, is much more difficult because of their bulk.

Teotihuacan is located close to two sources of obsidian, one of which was highly valued in Mesoamerica for its unusual green color. Early in its history, Teotihuacan merchants began trading their obsidian for other valuable goods from nearby centers. Obsidian was the steel of Mesoamerica; rulers wanted to amass as much of it as they could to use as weapons, ritual knives, and ornaments. As demand for Teotihuacan's obsidian grew, the city widened its export networks.

TEOTIHUACAN'S MERITOCRATIC MILITARY: Teotihuacan's wealth and the importance of the obsidian trade encouraged it to develop a professional military to increase its economic empire. At its height, at approximately 500 CE, Teotihuacan had an army that may have included as many as 151,500–334,563 men (in comparison, the U.S. sent only 250,000 troops to invade Iraq in 2003). Unlike previous Mesoamerican militaries, which are better characterized as "small groups of hit-and-run raiders," the Teotihuacan army was professional. As in modern armies, the Teotihuacanos divided their military into a number of formations armed with different weapons. Skilled soldiers on the front line wielded *atlatls*, a type of spear thrower with a long range. Behind them stood troops armed with spears and clubs, ready to engage in hand-to-hand combat, and commoners eager to provide protection with slingshots. Teotihuacano generals devised battle plans before each confrontation in order to command the thousands of people engaged in each battle. In the heat of battle, officers communicated with their troops through battle standards (distinctive feathered head-dresses), so that "by watching the progress of the standards, any soldier in the unit [could] tell whether to advance, fall back, or stand his ground."[14] Teotihuacan needed its army not to protect the city, which was too big for any other Mesoamerican group to conquer it easily, but to protect its merchants and open up trade routes. To motivate its soldiers to do the bidding of the state, Teotihuacan could tempt them with promises of booty and of social advancement.

As at Mycenae, market principles operated in some economic spheres at Teotihuacan, such as the market for luxury goods, whereas tradition and the state probably regulated other aspects of the economy, such as the production of corn. A large percentage of Teotihuacan's population was made up of merchants and artisans, in contrast to earlier urban societies, like Mesopotamia, where cities contained more farmers. Although we do not know precisely how the market at the center of Teotihuacan operated and how the city's government controlled, regulated, or stimulated Teotihuacan's economy, the archaeological evidence emphasizes the economic diversity of this city.

GOVERNMENT AND THE LAW

Dua-Khety worked for the state, while Pepy was set to embark upon his training at the Egyptian court. As we saw in the previous section, early governments employed large numbers of the first citizens. But how did these governments actually work? Were they monarchies, democracies, or something entirely different? Although researchers have spent an enormous amount of time studying the political aspect of ancient societies, a lot of their reconstructions of ancient politics have been hopelessly skewed. People have generally studied ancient government not in order to understand it, but in order to understand their own system of government. As a result, people have tended to oversimplify ancient politics. For most of the nineteenth and early twentieth century, people saw Mesopotamia as a perfect example of "oriental despotism"—a society where the ordinary people were slaves, but the kings were tyrants with vast powers. Popular European views about the backwardness of "Orientals" influenced this image of Mesopotamian government. By the late 1940s, the cold war instead inspired some American historians to suggest the opposite, that the first Mesopotamian government was not made up of power-hungry dictators, but of democratic councils. At the same time, universities emphasized courses on Western civilization that taught that the United States was the triumphant heir of 5,000 years of progress. Although neither of these extreme views are true, they do highlight important aspects of Mesopotamian politics. Mesopotamian government combined monarchy with tribal and democratic elements. Similarly, many archaeologists are now hotly debating who ruled the Indus civilization and how it was ruled. The first excavators in the Indus valley were British colonial officials; their histories of these Harappan cities were influenced both by colonialist notions of the Indians and by Hindu accounts of history. Now a new generation of Indian and foreign scholars are rejecting previous certainties that, like Mesopotamia or Egypt, the Indus valley was ruled by a powerful king. Instead, they suggest that the Indus cities had no real rulers.

MESOPOTAMIA: PRIMITIVE DEMOCRACY OR PRIMITIVE DICTATORSHIP?

ORIENTAL DESPOTISM AND EARLY EXCAVATIONS IN MESOPOTAMIA: When English and French adventurers first discovered the ancient Assyrian and Babylonian cities, people throughout Europe and North America were excited by their finds. In the nineteenth century, knowledge of the Bible was more widespread than it is today, and the idea of finding the treasures of ancient Nineveh, Babylon, or Ur of the Chaldees excited many. The prime minister of England, William Gladstone, attended the public translation of a broken tablet about an ancient Mesopotamian flood story similar to Noah's flood, and a major English newspaper sponsored an excavation to unearth the rest of it. Yet there was a downside to this public interest—much of what archaeologists found was interpreted not on its own terms, but in light of the Bible or more recent history. Because the Assyrians and Babylonians are evil tyrants in the Book of Kings, it is unsurprising that the first histories of Mesopotamia emphasized tyranny and absolute power. One of the first textbooks of Ancient Near Eastern history, published in 1869, sums up the general picture: "The Assyrian monarchy presented. . . . a type [that] we may still see in the Ottoman empire at Constantinople . . . as opposing . . . insurmountable obstacles to the progress of all liberty and civilization; an unbounded unrestrained despotism, interrupted from time to time only by revolutions plotted in the palace."[15]

MESOPOTAMIAN KINGSHIP: Several documents from Mesopotamia also seem to support this view. For historians in Mesopotamia, history was a long list of kings and dates, just as it was for their Egyptian contemporaries (Chapter 3). The Sumerian King List, for example, interprets early Mesopotamian history in terms of "kingship"—an institution which heaven granted to different cities—but only for a short time. Kingship also emerges as a central institution in a number of Sumerian myths, suggesting that many of the inhabitants of ancient Mesopotamia probably considered it—if not the only form of government, at least the best form of government. Additionally, from 2600 BCE onward, kings commissioned royal inscriptions, many of which still survive, that also help us to understand something about the role of kings in Mesopotamian society. Not all of these are particularly informative. An early inscription reads simply, "To (the god) Zababa, Uhub, prince of Kish, son of Puzuzu, conqueror of Hamazi, has dedicated (this vase)."[16] Luckily, later inscriptions usually make for more exciting (or at least more detailed reading), as kings brag about defeating their enemies in battle or building massive temples for their gods. Often these documents do not describe what really happened accurately at all. Royal inscriptions were a form of propaganda—kings put them up to convince the populace that they were doing a good job.

THE DEVELOPMENT OF KINGSHIP: Kingship is so central to Mesopotamian life that its late development may come as a surprise. The first Sumerian kings appear around 2600 BCE—almost 1,000 years after the appearance of the state in Mesopotamia. Who governed Mesopotamian cities before these kings? And why was kingship invented? As we saw in Chapter 3, the earliest rulers of Uruk were representatives of the gods, who had little personal power, at least when compared to the almost omnipotent bureaucracy that really governed this city. The only monumental buildings—and the center of administrative power—were a series of temple complexes. All Mesopotamians believed that their city belonged to the city's god, who was its ultimate ruler. The names of the oldest Sumerian cities were composed of two main elements—the cuneiform sign for city and the name of its god. Given this, it makes sense that the first leaders ruled as representatives of the gods. By 2600 BCE, however, this had changed, and the first palaces appear, often built far from the temples that dominated the central acropolises of ancient Mesopotamian towns. These early kings began their political careers as generals, chosen by the temple establishment or a group of city elders during periods of warfare. As they became stronger, they established a power base separate from the temples (which were still run by the priests, who now had little secular power). In time, some of these early kings were able to take over the temples, making themselves the main administrators of the houses of the gods, as well as of their palaces, thus reversing the original order. In the view of many, this resulted in a situation where "one individual, the ruler, united in his hands the chief political powers: legislative, judiciary, and executive."[17]

PRIMITIVE DEMOCRACY IN MESOPOTAMIA: But is this the whole story? Were the cities of Mesopotamia always ruled by despotic priests and kings? Did the people of these cities never have a voice of their own? Another thesis suggests that far from being primitive dictatorships, early Sumerian cities were governed according to a democratic model, an ancient ancestor of the assemblies of classical Athens.[18] Sumerian myths are the oldest available sources that discuss political organization. Although they literally describe how the gods govern themselves, they may also reveal insights into how Sumerian government worked. The myths suggest that a council, composed of all citizens, ruled each Sumerian city. During times of crisis, the council assembled to resolve it. It could either act as a court to settle legal disputes, nominate a "priest" to handle administrative matters (like organizing a group of men to fix a broken canal), or nominate a "king" to serve as a war leader. Both of these officials were appointed only for the duration of the crisis. The assembly model worked both for individual cities and for Sumer as a whole. A number of seal impressions from around 2900 BCE, before the appearance of kings or palaces, show a list of city names, which may be evidence of an ancient league of cities. Similarly, later documents, from around

2400 BCE, also refer to a confederation of cities—called the Kiengi league. Representatives of these cities may have assembled whenever a crisis threatened the whole of southern Mesopotamia. It took a long time (at least a millennium), and a lot of luck, for individual kings to create a permanent institution of kingship. In this scenario, monarchy followed democracy.

POPULAR ASSEMBLIES IN LATER MESOPOTAMIA: Many historians doubt the validity of this reconstruction. They argue that just because stories describe the gods acting in a certain way, it does not follow that people actually acted that way in the distant past as well. Myths do not always, or even often, preserve historical truths. It would be folly to argue that the first Romans gave their children to wolves to bring up, just because a wolf supposedly mothered the founder of Rome. Rather than accepting that the most ancient form of Mesopotamian government was the most democratic, some historians argue the opposite. We have much more information about the operation of city assemblies during later periods of Mesopotamian history—particularly after 1800 BCE—than we do for the distant period between 3000–2600 BCE. During this period, citizens gained increasing political power, until by about 650 BCE, the citizens in several cities in Babylonia could make decisions independently from the king. The king had no right to draft these citizens into the army, to tax them, or to impose legal decisions upon them. Legal texts from several cities discuss the meeting and decisions made in assemblies. Just like in America in the twenty-first century, Mesopotamians had the right to trial by a jury made up of their peers—other residents of their city. Sometimes these assemblies seem even more democratic than the Athenian ones, which were limited to free, land-owning men. Mesopotamian assemblies were made up of a wider slice of society, including manual laborers and probably women. These assemblies protected the people from a king intent on abusing his power. As a result, it seems that "the citizenry itself was most often in charge" in Mesopotamian cities.[19]

TRIBAL GOVERNMENT: Mesopotamian society may have been quite urban, but its countryside always contained nomadic tribes in addition to farming villages. Just like in modern Iraq, where tribal leaders have enormous influence, nomads in Mesopotamia were an integral part of the political landscape. The importance of tribes generally increased when that of the state decreased. From about 2000–1800 BCE, for example, Amorite tribal leaders gained power throughout western Asia, from Aleppo in Syria to Larsa in southern Iraq. At first, these tribes represented a threat to the Sumerian cities; the most important event in the reign of one Sumerian king was the construction of a wall to keep out these hostile nomads. Over time, however, Amorite chiefs became urban rulers. Hammurabi (1792–1750 BCE), the king of Babylon who created an empire in Mesopotamia which lasted almost 200 years, is the descendant of a line of Amorite chiefs. The later kings of Assyria, who similarly created empires which, at their height, incorporated the area from Egypt to Eastern Iran, also styled themselves as the descendants of nomadic kings, "kings who lived in tents."[20] Even after tribal chiefs became urban kings, tribes often maintained independent power. The kings of Mari, a city on the Euphrates, who were themselves tribal leaders, signed mutual defense agreements with myriad nomadic tribes. Just as in urban society, one chief did not necessarily dominate tribal society. Instead, tribal councils could challenge the actions of a chief.

Over the 3,000 years of Mesopotamian history, many powerful kings arose. Some of them ruled just one city-state, while others were able to conquer vast empires. Epics, statues, rock carvings, and even drinking songs celebrate the exploits of these rulers, but their reigns are only a small part of the story of Mesopotamian government. Nomadic chiefs, tribal councils, and urban assemblies also contributed to the government of Mesopotamia. Sometimes law codes and legal cases, administrative receipts, and diplomatic correspondence record their activities, but often we have only a vague reference to their influence. Nevertheless, despite Mesopotamia's own emphasis on the importance of the king, politics in Mesopotamia involved the entire people, both in the cities and the countryside.

INDUS VALLEY: WHO NEEDS A STATE ANYWAY?

In 1826, a British army deserter in India found scattered bricks, fragmentary walls, and "a ruinous brick castle" in a small town called Harappa. His descriptions inspired some British colonial officials and explorers to excavate the mound. They thought that the ruins at Harappa belonged to a thriving city described by Chinese Buddhist pilgrims in the early centuries CE. Further excavations revealed that the vast remains at Harappa were nearly 3,000 years older and represented the earliest civilization on the Indian subcontinent.

In some ways, the initial confusion over Harappa underlines a major problem in our interpretations of this society. We have attempted to understand the Harrapan cities inappropriately, either in terms of other early societies (particularly Mesopotamia, given the close trading contacts between the two civilizations), or in terms of Hindu traditions. Hindu historical sources argue that Hindu society emerged sometime during the second millennium BCE, with the destruction of the Harappans. The founders of this society were supposedly a group of horse nomads, known as the Aryans, who conquered and enslaved the peaceful farmers and craftsmen who had built Harappa (for more on the Aryans see Chapter 2). As a result, most scholars have thought of the Harappans as a dead end, a civilization that disappeared and had little impact on later Indian history.

THE INDUS VALLEY CITIES: At Harappa and the other main capital of this civilization, Mohenjo-Daro, years of excavation revealed large, richly built houses, complete with indoor plumbing and luxurious bathrooms, dating to approximately 2600–1900 BCE. These houses were laid out along straight, paved streets, and included private wells. Underneath the streets, a system of pipes drained away wastewater, ensuring that Mohenjo-Daro was much cleaner than most ancient cities (the streets of Uruk would have been filled with trash and animal and human waste, much the way the streets of Paris or London were, even in the eighteenth century CE). Evidence from Harappan cemeteries shows that the inhabitants of the Indus Valley benefited from such enlightened public health initiatives. Analysis of their skeletons reveal that they

FIGURE 7
This aerial view of the excavations at Mohenjo-Daro illustrates the careful planning and precise layout of the city.

were generally healthy, without the stress due to poverty and malnutrition that mark the bones of the inhabitants of other early states. Artifacts like carnelian necklaces and painted terra-cotta rattles demonstrate that the inhabitants of this city were skilled craftsmen. As in Mesopotamia, carefully maintained irrigation channels watered the wheat and barley crops, ensuring abundant harvests to feed Mohenjo-Daro's citizens.

A WORLD WITHOUT RULERS?   This picture is clearly different from that of life in Mesopotamia, Mycenae, or Teotihuacan, for one key reason: all of the archaeological evidence we have reflects the lives of the ordinary people, not the rulers. The buildings that we have described are simple houses, not palaces or temples. The art of the Indus valley usually shows animals: begging dogs and playful monkeys. When human figurines do appear, they are generally faceless, unlike the elaborate sculptures of Mesopotamian kings. In the Indus valley, 150 years of archaeological research has revealed no undisputed palaces, temples, kings, or warriors. The Harappans also did not distinguish between elites and commoners after death, but buried everyone in a similar fashion, with few grave goods. Certainly there are no Harappan pyramids, or even elaborately furnished graves like in the Royal Cemetery of Ur in Mesopotamia. The finds that we do have, along with these important absences, suggest that Indus society was organized differently from other ancient civilizations. Unlike them, it was not a state. Although there is clear evidence for economic specialization and surplus production, there is no evidence for a political hierarchy or for a centralized administration.

Despite the lack of archaeological evidence for state institutions in the Indus Valley, many writers have tried to make the Harappan data fit the state model. The state is the only complex political form that exists currently, and it seems impossible that writing, urbanism, and a society differentiated into several professions could be anything other than a state. Even if there is no evidence for palaces, temples, or rulers, most historians have argued that given their complexity and cultural uniformity, Harappa and Mohenjo-Daro must have been the major cities of a large state. They have claimed that the absence of archaeological evidence of kings does not prove their absence. Even in later periods in India, when the Vedas and other historical sources clearly indicate the presence of powerful rulers, kings left little mark on the landscape. They never built large statues of themselves nor erected royal inscriptions boasting of their deeds. It is not until the Mughal rulers of the sixteenth century CE that Indian kings devoted time and resources to self-aggrandizement. Yet despite this tradition of understated kingship, there is clear evidence for

kings, princes, and nobles in later Indian history and archaeology—
evidence that is lacking for the Indus valley cities.

ALTERNATIVES TO THE STATE: Since class hierarchies, powerful kings, and
priests did not exist in the Indus civilization, how do we explain its political
organization? What is the alternative to the state? If we accept the evidence
that Indus civilization was every bit as complex as other ancient state-based
civilizations of the same period (such as Egypt and Mesopotamia), we must
propose some mechanism that can integrate a complex society in the absence of the state. The Indus civilization covered an area of roughly 1,210,000
square kilometers, more than a dozen times greater than the area of Egypt
and southern Mesopotamia combined, and more than twice the size of
Texas, for at least 600 years. Its inhabitants used similar city-planning techniques, manufactured similar goods, and used the same writing and mea-
surement systems over this entire area. Something must have served to
assimilate each city into this cultural milieu. Perhaps the best way to understand Indus civilization is to consider the operation of a powerful nonstate
institution in modern India: the caste system.

THE HARAPPAN ROOTS OF THE CASTE SYSTEM: The caste system under-
lies traditional Hindu society. There are a number of castes that are defined by kinship and profession. People are born into castes and they
marry within their castes. Castes are also ranked hierarchically, based on
a scale of purity-impurity, with the two extremes represented by the pure
Brahmans—a priestly caste—and the untouchables—whose professions
link them with sources of pollution, like hair, feces, menstrual blood, or
garbage.[21] The caste system is remarkably flexible and has adapted to
changes in society throughout history. Between the two extremes of
Brahmans and untouchables, all the other castes are ranked relative to
each other, but their comparative position may change over time, and
castes may be created or destroyed, as marriage practices change. Because castes are traditionally linked to different professions, every village
or city includes several different ones. Because men do not marry women
from other castes, they must choose brides from other communities. This
exchange of brides unites distant communities and introduces each community to ideas that are common in the outside world. Another interesting
aspect of the caste system is the position of the king; political power and
rank are separate. The king does not occupy the top position in the caste
system; instead, he is spiritually subordinate to the priests. At the same
time, the priests, although possessing higher rank, are in reality subject to

the king. Hindu priests approach the gods through sacrifice, thus helping to ensure the purity of the king and of his kingdom.

RECONSIDERING THE ARYAN CONQUEST: Historians have generally dated the emergence of the caste system to approximately 1700 BCE, when the Aryans conquered Mohenjo-Daro, turning its occupants into slaves (the lowest castes) and divided themselves into the three top castes. Now, however, few archaeologists believe that the Aryans had anything to do with the destruction of the Harappan civilization. There is no evidence for military conquest of any Harappan cities. If the Aryans did "arrive" during the second millennium BCE—and were not indigenous to India—their arrival probably occurred after the Harappan civilization had already collapsed. The caste system today operates most widely among settled farmers, which suggests that the system developed among farmers, like the Harappans, or pre-Harappans, not foreign pastoralists. The Aryan innovation may have been the insertion of a class of pastoralists, and the imposition of a leader, the king, who has an uneasy relationship with the Brahmans.

THE HARAPPAN LEGACY: Of course, the caste system has been transformed immeasurably since the second millennium BCE, so much so that it might seem foolish to suggest any connection at all. Yet certain elements in modern Indian and ancient Harappan caste societies are parallel. Although we have no statues of gods, or distinct temples from any Harappan centers, we do have an obsession with purity, seen in the baths in Harappan houses, the great baths in Mohenjo-Daro, and the Harappan sewage system. There are other signs of continuity between Harappan practices and later Hinduism. Harappan seals show the sacrifice of a water buffalo, a ritual which is associated today with the goddess Durga, and suggests a more general Harappan emphasis on sacrifice. Other seals show a many-faced, horned god, surrounded by animals, perhaps the god Shiva. Finally, there may be some archaeological evidence that different classes—perhaps subcastes distinguished by profession and marriage practices—lived in separate parts of Indus cities. At Kalibangan, a medium-sized town, the excavator believes that the priests, merchants, cultivators, and artisans all lived in distinct quarters, physically separated from one another.[22] Bangles and seals, two common artifacts in Indus valley houses and graves, may also have served as caste-identifying marks. The way women wear bangles in the Indian subcontinent today communicates social status to others; the same might have been true for the Indus valley. Both men and women may also have used seals with animal motifs, thousands of which have been found in Indus settlements, to indicate their position in the Indus system. By carrying a common unicorn seal, for example, you may have advertised your allegiance to a cultivator caste, whereas only a few members of the priestly caste may have owned the rare zebu bull seals.[23] Several of these seals are inscribed, suggesting that Indus writing may have been used to affirm this system.

THE EMERGENCE OF CASTE: How could a castelike system emerge in the Indus valley? Its roots may lie in some of the earliest agricultural villages. As we have seen, many village societies are based on kinship; you might live in a village because your ancestor founded it long ago. When people began specializing in different occupations, professions were organized according to families. We can see that today in many English names—like Smith, Potter, Cooper, and Weaver—family names that originally referred to family trades. In a caste system, certain lineages, special branches of an original family, may have adopted certain professions. Over time, their profession became more important to their identity than their original clan. People practicing these professions in different villages probably shared a similar economic position, and similar ritual practices—tied up with their professions—quickly emerged. When people began marrying within their professional set, rather than merely in their kinship group, the caste system was born. The practice of marrying within your caste but outside of your village explains the incredible uniformity of Harappan society over a vast distance. The women who traveled to distant villages as brides brought with them a dowry of information. As castes grew up, a remarkably stable world emerged.

A caste system may be an alternative to the state. It serves to integrate a society, and like the state, it uses hierarchy as the main tool to do so. Unlike the state, however, a caste system does not rely on an authoritative center. It is difficult to understand how a caste system without a state may have worked, because it is so foreign to our understanding of government. It may help to compare the Indus caste system with a Western market economy. Western market democracies operate under the assumption that the market can set fair prices for both products and labor, due to the interaction of supply and demand. The market forces that serve to regulate the economy are invisible; when we name them, we are using a shorthand term for a system that arises out of the interaction of millions of people who calculate what and how much they will produce and buy according to their own interests. The interaction of the proto-castes in the Indus valley system, each performing slightly different tasks in different ways, may have created a homogenous, well-integrated society.

THE DRYING-UP OF THE SARASVATI AND THE COLLAPSE OF THE INDUS CIVILIZATION: Why did this system, which was stable enough to last 600 years, collapse suddenly around 1700 BCE? Some scholars suggest that the roots of its destruction lay in its unusual nature. Whatever the organizing principle behind Indus society, it failed, and no other complex, nonstate society ever arose anywhere in the world. Or its end may have been due to environmental factors, like a worldwide drought that caused a drastic change in the course of the rivers that allowed these cities to bloom in the midst of a harsh desert. The majority of settlements belonging to the Indus civilization lay not on the Indus River, but on another river, the Ghaggar-Hakra, also called the Sarasvati. Today, the Sarasvati is a tiny tributary of the Indus, but there are signs that 5,000 years ago, it contained more water and was independent of the Indus. The Sarasvati began drying up following a series of earthquakes that caused its tributaries to flow into the Indus. As Indus flow increased, the path of that river changed, flooding nearby settlements. The disappearance of the Sarasvati and the devastating floods of the Indus were too much for the Harappans. Settlements were abandoned as each river failed, until the entire population dispersed into villages in areas that were not affected. An echo of the Sarasvati's ancient importance to the Harappans survives today in the worship of the Hindu goddess Sarasvati, who guarantees prosperity and abundance, a role that the Sarasvati river may once have fulfilled as well.

CHOLERA AND COLLAPSE IN THE INDUS VALLEY: Another thesis suggests, ironically, that the Harappan obsession with purity—and specifically their elaborate plumbing system—doomed them in the end. Most Harappans drew their drinking water from a system of open wells, many of which were situated close to the drains for wastewater. This meant that it was easy for sewage to contaminate drinking water, allowing for the spread of infectious diseases, like the water-borne disease cholera. Cholera can cause severe dehydration and death within only a few hours. It spread to the rest of the world during the nineteenth century, but an ancient Vedic medical text describes cases occurring in India in the second millennium BCE. Indeed, the face of Kali, the goddess of plagues and destruction, may represent the visage of a cholera victim, whose skin turns mottled and dusky from severe dehydration. In times of epidemic, city dwellers often flee to the countryside to escape the plague. This is an effective response to cholera, as dispersing the population decreases the likelihood of further infections. Every attempt to recolonize the cities, however, and use the same water and waste system, may have met with similar fatal outbreaks, until the Harappans gave up on cities altogether.

The place of Mohenjo-Daro and Harappa in Indian history has shifted subtly over the last century. At first, the civilization that these two cities represented was seen as a sidetrack to Indian history, a precursor that died out utterly—destroyed by the Aryans that created proper Indian culture. When we reconsider the Aryan destruction of this civilization, however, we can see how many of the key elements of Indus civilization have continued into more recent times. The archaeologist Douglas Price nicely sums them up: "ceremonial bathing, ritual burning, specific body positions (such as the yogic position) on seals, the important symbolic roles of bulls and elephants, decorative arrangements of multiple bangles and necklaces (evident from graves and realistic figurines), and certain distinctive headgear—all are important attributes of ancient Harappan society that remain at the heart of contemporary Hinduism."[24] Despite this, there are many fundamental differences between later Indian history and the Harappan experience. We cannot be sure precisely how Harappan society functioned, but we do know that the techniques they used were different from those of any other complex society in the world. We still have much to learn about how a civilization can be run without a formal government.

CONCLUSION

The origin of the state was the first step toward producing the sorts of societies we live in today. The 2000 U.S. census shows that 80.4 percent of Americans have graduated from high school, while less than 1 percent of them work as farmers.[25] These statistics are very different from those experienced by ancient societies. It is hard to imagine that more than 5 percent of people living in Mesopotamia during the approximately 3,000 years when cuneiform was the dominant script could read or write, while none of them had completed 12 or more years of education. On the other hand, a much greater percentage of the population had to take time out to sow and harvest crops, even if they also worked at other tasks. The same is true for every other early civilization. Similarly, modern conceptions of government and law anywhere in the modern world, with their reliance on written codes of law, are far removed from the more organic situation that prevailed in the Indus valley. Nevertheless, many of the details of our world are simply elaborations on an ancient model, one which was worked out slightly differently in different parts of the world.

TIMELINE

Region	6,000 BCE	5,000 BCE	4,000 BCE	3,000 BCE	2,000 BCE	1,000 BCE	0 BCE	1,000 CE	2,000 CE
West Asia		Tokens: a precursor to writing	First cuneiform writing Division of labor						
East Asia		Potter's marks in China		Complex chiefdoms	First chinese writing (Zhou dynasty) State in China				
South Asia				Indus Valley civilization	Collapse of the Indus Valley civilization				
Central Asia									
Southeast Asia									
Australia									
Europe					Traders at Mycenae				
North Africa			First Egyptian hieroglyphics						
West Africa									
East Africa									
South Africa									
North America									
Meso-america					Earliest Zapotec writing	Zapotec and Olmec writing Mesoamerican calendar system	Teotihuacan Professional armies in Mesoamerica	Tenochtitlan and the Aztec Empire	
South America							Quipu system		

CHAPTER FIVE

ANCIENT WORLDS

GETTING STARTED ON CHAPTER FIVE: This chapter explores the emergence of an African-Eurasian "world"—a collection of interrelated societies and civilizations—beginning at 4000 BCE and considers whether a similar network existed in the Americas, uniting the Andes, Mesoamerica, and North America. It concludes by considering the collapse of an ancient network that comprised states, chiefdoms, and village societies. How might the emergence of the state in the Near East have led to changes in European chiefdoms? What roles did pastoral nomads play in the interaction of state societies? Why have historians so rarely recognized pre-Columbian American networks (and ancient networks in general)? What were the benefits and disadvantages to the establishment of these world societies?

COMPARISONS AND CONNECTIONS

World historians look at past events through two frameworks: comparison and connections. In general, we have compared and contrasted transitions in different parts of the world, looking at how people in Mexico, Africa, and southwest Asia all domesticated plants and animals, for instance, or the similarities and differences in the invention of writing in southern Iraq (Mesopotamia) and Veracruz, Mexico. Except in restricted geographical areas, we have not stressed connections. For the most part, that has been because they have not existed, at least not on a global level. It also reflects a prejudice in archaeology against explaining the past solely through contact. Up until the 1930s, archaeologists saw everything in terms of connections. Any innovation was thought to be a sign of invasion (denying "natural" intelligence to most groups), and archaeologists confidently spoke of different "peoples" and ethnicities in prehistory. In the 1930s and 1940s, however, the German government used evidence drawn from excavation and survey to justify their conquest of eastern Europe. They claimed that these areas belonged (pre)historically to a greater Germany: their current inhabitants were just degenerate Slavs, ethnically inferior to the Teutonic warriors who had once tamed the land. In reaction to this, during much of the last 60 years, archaeologists have shied away from discussions of invasions, contacts, and the spread of ideas in early history.

In contrast, when we talk about current events, or modern history (since 1500 CE), we generally stress connections. Like the proverbial butterfly in Brazil, the flapping of whose wings causes a hurricane in Florida, we now recognize that events in one part of the world may have far-reaching consequences in others. Military decisions made by an ambitious Mongol prince, Chinggis Khan, were one factor in the spread of the Black Death in Europe and the end of feudalism.[1] Decisions made by a Genoese adventurer in 1492 hoping to sail to China, for example, were one factor in the collapse of the Aztec empire 30 years later. The world wars of the twentieth century, fought primarily in Eurasia, contributed to the fall of European colonialism in Africa.

But world historians have increasingly argued that connections among human societies operated even in ancient times and were significant in shaping the histories of ancient peoples. One prominent world historian put it like this:

> In the ancient Middle East, the resulting interactions among peoples living in different landscapes, with diverse languages and other outward signs of civilized diversity, led to the emergence of a cosmopolitan world system between 1700 and 500 BCE. . . . There is a sense, indeed, in which the rise of civilization in the Aegean (later Mediterranean) coast lands and in India after 1500 BCE were and remained part of the emergent world system centered on the Middle East. . . . One may, perhaps, assume that a similar primacy for economic exchanges existed also in earlier times all the way back [to] the earliest beginnings of civilization in ancient Mesopotamia.[2]

If we begin to look at ancient history not only through the lens of each individual civilization, but also with a wider view that embraces the many connections among them, we may be able to learn more about how history actually operated. In this chapter, we will examine several types of contact among human communities in the ancient world. We will begin with the spread of technology and luxury goods in Eurasia and Africa from 4000–2500 BCE and its effects on a wide range of societies, including states, chiefdoms, and pastoral societies. Second, we will consider whether a similar network of relationships existed between Mesoamerica and North America from 800–1350 CE. Finally, we will consider what happened when an ancient network collapsed in the Mediterranean region during the twelfth century BCE.

The Secondary Products Revolution

Down-the-Line Trade and the Establishment of World Systems: If we are ever going to satisfactorily answer how states and complex chiefdoms emerged in Asia, northern and eastern Africa, and Europe, we must consider this enormous area as one zone of interaction. The emergence of at least three complex societies at roughly the same time suggests that they were not simply responding to local environmental conditions or population pressure. Instead, one of the forces favoring their emergence was probably their interaction. Exactly how these states and chiefdoms interacted is difficult to determine. The world from 3500–1500 BCE, when this early "world-system" appeared, worked very differently than it did later. We cannot simply assume that relations between farming communities in Bulgaria and Egypt were anything like later trading relations between these places. Although trade did exist, it was not on a large scale

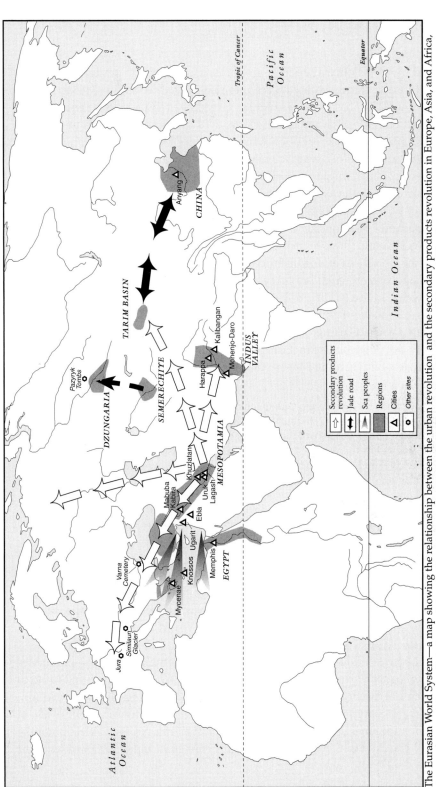

The Eurasian World System—a map showing the relationship between the urban revolution and the secondary products revolution in Europe, Asia, and Africa, the migrations of the Sea Peoples, and the relationship between Central Asian nomads and sedentary Chinese civilization.

and involved only a few commodities. Moreover, this trade was usually not direct—Indus valley merchants may have traveled to Mesopotamia, but they did not go on to Anatolia, Greece, or Hungary. Instead, valuable products, often luxury goods, spread from one community to the next through a network of gift exchange. A trader from Habuba Kabira, Syria, might give a chief living 300 kilometers upstream a fine set of silver wine glasses, which might then be included in his daughter's dowry when she married the ruler who lived over the next mountain. Over a few hundred years (a very short time archaeologically), that wine service could travel thousands of kilometers. Some of the concepts associated with it could travel even further. A visitor from the Balkans, impressed with the beauty of the fluted silver, for instance, might commission a copy of it from the village potter—a cheap imitation of the metal vessels in clay. Another traveler, invited to drink wine out of these cups, might marvel at the wonderful effects of this exotic drink and try to brew something similar at home, perhaps using barley and blackberries rather than grapes. Andrew Sherratt, a specialist in European prehistory, explains how these early contacts between the Mesopotamian center and the European backwater worked:

> Within this sequence, bulk exchanges and direct commercial engagement with the "civilized" world are fairly late features of European prehistory. . . . Before that, it was the indirect ideological and technological influence (the two are not really opposites, but complements!) that penetrated into the temperate parts of the continent, carried often in quest of things which were remote, valuable and useless—like gold and amber—and bringing knowledge of new luxuries—like alcohol, wool and bronze. . . . From 3500–2400 BCE, innovations of general desirability spread across Europe over a broad front, creating new and enlarged cultural units . . . the effects of Near Eastern innovations were no less real for being transmitted indirectly.[3]

We can begin to piece together some of this complex story by looking at how new technologies (agricultural intensification and metallurgy) and luxury goods (especially drugs like wine, beer, hemp, and opium) were adopted and adapted across the old world, funding chiefdoms and states, and how the contact between states, chiefdoms, villages, and sometimes hunter-gatherers both created and maintained the powers of new leaders, from "big men" to kings. In many cases, it is impossible to decide exactly where certain inventions, including the cart, plough, horse riding, and metallurgy, originated, because we find them contemporaneously in places like the Ukraine, Iran, southern Mesopotamia, and the Balkans. What is more significant than the exact place of their origins, "is the extent of the interconnections suggested by the huge area over which, for example, wheeled vehicles appear—seemingly simultaneously."[4]

THE SPREAD OF TECHNOLOGY

THE SECONDARY PRODUCTS REVOLUTION: In the Near East, the rise of the state stimulated the invention and adoption of several technological advances related to farming. By approximately 3500 BCE we have evidence for a slew of inventions that increased agricultural yields, including the plough, the wheel, the domestication of draught animals, and the development of milk and wool production. Improvements in agricultural efficiency meant that fewer farmers could produce more food, freeing up more people to work on other things, like building pyramids or brewing beer. As a result, it was in the best interests of ambitious chiefs, kings, or priests to invest in promising new technologies. Andrew Sherratt refers to the invention and spread of this agricultural complex as the "secondary products revolution"[5] in order to highlight the fact that many of these improvements were simply finding new uses for animals that had been domesticated millennia before. Recognizing that animals could supply food, clothes, and labor presented new possibilities. These innovations fueled both emerging states and chiefdoms, opened up new land for settlement, and allowed for the development of new ways of life, like pastoralism and commerce. In Mesopotamia, Egypt, the Indus Valley, and China, they permitted the establishment of class-based societies. In Europe and central Asia, on the other hand, this development led to radically different lifestyles and societies—transforming where and how people lived, relations between men and women, and chiefly politics.

THE DEVELOPMENT OF MILKING AND PASTORALISM: It might seem hard to believe, but prior to about 4000 BCE there is little evidence for milking animals; instead, most domesticated animals were raised for meat.[6] Archaeologists can tell whether animals were raised for meat or milk by comparing the ages of the animals at death. If you are raising goats, sheep, or cattle only for meat, it makes sense to slaughter them when they are still young and tender. On the other hand, if you are raising goats or cows for milk, it is better to keep female animals for longer. In the Levant, at approximately 4000 BCE, analysis of animal bones suggests that suddenly, people were keeping sheep and goat mostly for milk, not mutton. At the same time, pottery vessels, which could have been used as butter churns, suddenly appear in houses. So why is the development of milking (and the likely invention of yogurt, cheese, and butter) so important? First, using animals for milk is more cost effective than raising them for meat. Animals are expensive. Although they can graze part of the year, they often need to be fed stored grain during the winter, which could be used instead to feed people. If you're just raising them for meat, all of this care and feed is wasted until you slaughter them and enjoy your feast of lamb chops. On the other hand, if you keep animals for milk, the food you feed them is quickly converted into a nutritious, protein-rich food for you. The development of milking allowed people to exploit new environments, which were not suited to planting, but were suited to grazing. Before milking, pastoralism—herding sheep, goat, cattle, or horses full-time—did not exist. It did not make sense to herd animals in hostile environments if the only source of food was meat from those animals; the herd would disappear too quickly and the shepherd would starve to death. As a result, only villagers kept animals; they ate bread to survive and just enjoyed meat on feast days. On the other hand, many pastoralists live quite happily for months just on milk products. This allows them to range widely over land that is too dry for agriculture, and only occasionally visit villages to trade yogurt for bread, because no pastoralist lives without some contact with farming societies.

BREEDING WOOL SHEEP: At about the same time that people began consistently using animals (mostly goats) for milk, others began breeding sheep for wool. We have already reviewed the evidence for the development of wooly sheep in Uruk in the fourth millennium BCE (Chapter 4). Soon afterward, analyses of sheep bones show that people outside of Mesopotamia were letting sheep live longer, probably to exploit them for wool. The earliest wool in Europe comes from the French Jura and dates to about 2900 BCE. This was probably just about when it was introduced, as the Alpine Iceman, whose body was preserved in the Similaun glacier at about 3300 BCE, wore clothing made of animal skins and a woven straw cape. Throughout history, beautiful cloth has fueled trade routes. From the "Silk Road" to Paris fashions, people have been willing to pay fabulous prices for clothing. Unlike linen or leather, wool can be dyed easily and woven into intricate patterns. To someone used to wearing dull-colored animal skins and straw capes, brightly colored woolen tunics were probably a revelation, an instantly desired object that increased their wearer's beauty and his or her status.

THE INVENTION OF THE PLOUGH: Other inventions, like the plough and the cart, encouraged farmers to use animals for their labor. For the first five millennia of agriculture, the basic agricultural tool was the hoe, which farmers used to prepare fields for planting. In Mesopotamia, however, the plough[7] seems to have been first used around 5,000 BCE—based on plough marks preserved in Khuzistan, Iran. Twelve hundred years later, it had spread to Europe, where plough marks have been found under burial mounds in Denmark and Poland dated to 3800 BCE. Using a hoe, a woman can easily turn over enough land to provide herself, her husband, and her children with grain—but producing a surplus requires a lot more work. Hitch an ox to a simple plough, however, and that ox can turn over as much land as four women. Furthermore, the plough does a better job of preparing the soil and can be used on more types of land than the hoe. In Mesopotamia, it increased yields along with newly invented irrigation. In Europe, it allowed farmers to move away from their small patches of easily worked soil and colonize great tracts of forestland. As European farmers chopped down forests to make way for fields, they also created new pasture. In this way, the plough and pastoralism developed together as two aspects of the same revolution.

DOMESTICATING DRAUGHT ANIMALS: At roughly the same time that the Mesopotamians began supplementing human labor with animal power, people in Egypt and the central Asian steppes began domesticating animals solely for the labor they could provide. The donkey and the horse were domesticated in Egypt and Russia respectively in the fourth millennium BCE. Horses spread into the Middle East and Western Europe in the third millennium BCE, but were very expensive (because they had to be fed higher-quality food than other animals), and only became important in the second millennium BCE with the invention of the light chariot. At that point, the horse and chariot represented an enormous military advance. Donkeys, on the other hand, became immediately popular as pack animals in Egypt, where no one bothered to adopt the cart until 1800 BCE, 3,000 years after it was invented in nearby Mesopotamia, because donkeys and boats were so useful. Unlike horses, donkeys are very hardy and don't need much food. In Mesopotamia, where they may have been separately domesticated, they were used instead of oxen, to pull carts and ploughs and as caravan animals.

THE INVENTION OF THE WHEEL: The final aspect of the secondary products revolution was the invention of the wheel—and of carts and chariots. "Inventing the wheel" has become shorthand in popular culture for something primitive and essential. In reality, the wheel is far from essential. It was never used for transportation in the Americas, for example, not because people living in Mesoamerica or the Andes were too stupid to think it up, but because it was useless to them. In Mesoamerica, there were no domestic animals to pull a cart because the only domesticate was the turkey; whereas in the Andes, where domestic llamas and alpacas could have hauled one, the mountainous terrain prevented wheeled vehicles from being much use (cars have a hard enough time of it nowadays). The wheel was invented 4,000 years after the development of agriculture, probably about 5000 BCE somewhere in southeast Europe, west Asia, or central Asia. Like ploughs, carts were originally used as agricultural tools; they provided tenant farmers with a way to deliver their excess grain to the newly built Eridu temple. Wheels had an important secondary use as well; they could be used to mass produce pottery. Like textile factories, potters' wheels were ancient Mesopotamia's answer to the industrial revolution. The slow potter's wheel appeared alongside the plough and accelerated pottery production, but the true potter's wheel was invented only a thousand years later, at the same time that the state appeared in Mesopotamia.

THE DEVELOPMENT OF BRONZE METALLURGY: There is one other important technological advance that occurred at about the same time as the secondary products revolution—the invention of metallurgy. For a long time, archaeologists believed that the most important technology that spread in the fourth and third millennium BCE was the material whose name we give to these two "ages": copper and bronze. It seemed likely that skilled Uruk smiths invented metalworking techniques, just as they invented so much else. Yet new archaeological research has produced some surprising finds. The earliest copper and gold tools and ornaments have been found in the backwater of Varna cemetery, Bulgaria—not the urban centers of Mesopotamia. These coppersmiths did not make swords or axes or even knives, but cult objects and jewelry. With the rise of states, however, urban elites were quick to use this new product. Thousands of tablets found in Ebla, in northern Syria, record deliveries of metal to craftsmen, and of weapons, silver rings, and even bronze hoes to citizens. Metallurgy traveled across Eurasia along with the secondary products revolution.

THE SPREAD OF LUXURY GOODS

CONSPICUOUS CONSUMPTION: Of course, not all of the products introduced in the mid-fourth millennium could be used to streamline agriculture or to serve other strictly utilitarian purposes. Although an Anatolian chief may have been impressed with and interested in a newly invented plough, he would be just as excited by his first taste of beer, or a newly welded gold ring. Even today, we spend a lot on diamonds from Botswana, cosmetics from France, and vodka from Russia. People flaunt their wealth and importance through the goods they consume—a process that Thorstein Veblen famously termed "conspicuous consumption." There is no reason to believe that villagers in fourth millennium Armenia would be any less appreciative of alcohol or jewelry than we are today, or any less ready to use it to exhibit their own status. We can trace the spread of luxury goods like wine, hemp, and opium by looking at the spread of cups, censers, and other pottery containers. In many cases, the spread of drugs and alcohol accompanied the spread of utilitarian goods, while the combined effects of both innovations led to cultural transformation across Eurasia.

Very little of past life survives in the archaeological record. When people die, languages, stories, and traditions die with them. Clothing, blankets, wooden furniture, and food all rot or disintegrate. One material that archaeologists have in abundance, however, is pottery. People may break plates, but those broken pieces are practically indestructible and can survive millennia. As a result, archaeologists often characterize peoples by their pottery. Yet pottery "is only part of the language of cuisine. While the pots may have been admired, it was on their contents that the hospitality was judged."[8] Often a new type of food or beverage arrives complete with the containers and utensils used to serve it. When the English began importing tea from China in the seventeenth century, they also imported "china" teacups and teapots in which to drink it.

THE INVENTION AND SPREAD OF ALCOHOL: The same was true in the fourth millennium BCE. Residue from an Iranian jug dated to 3000 BCE shows that someone in the community was drinking wine. Mesopotamian art shows people consuming beer, while the epic of Gilgamesh celebrates the sound counsel of a barmaid who tells Gilgamesh to enjoy life while it lasts. The appearance of several distinctive styles of cups throughout Europe soon after the invention of alcohol in the Middle East may illustrate alcohol's rapid spread. After about 3500 BCE, pottery copies of metal drinking sets appear in Hungary, and abundantly decorated goblets spring up in Britain. Analyses of the residue within shows that the farmers in this area were enjoying an occasional drop of a distinctive alcoholic drink, something like beer, but flavored with honey and wild fruit. These farmers seem to have copied Mediterranean habits—substituting beer for wine and pottery for metal containers. There is no archaeological evidence for alcohol before this period, and it seems likely that a drinking culture spread across Europe from the Middle East, along with other inventions, like the wheel and the plough.

SMOKING CULTURES OF THE STEPPE: In central Asia, lack of easy access to grain or sweet fruits made it difficult to brew beer or produce wine. At some point, probably during the first millennium CE, some steppe people invented koumiss—fermented mare's milk—as a substitute for wine. Before that time, these sheep and goatherders probably amused themselves with smoking rather than drinking. The Pazyryk tombs of Scythian warriors in Siberia contain cannabis seeds, along with bronze censers and other paraphernalia. People in Western China, in Xinjiang province were also buried with ephedra, another stimulant which is smoked, while in other central Asian tombs, people were buried with bags containing hemp, opium, or ephedra. Archaeologists find abundant censers—containers used to burn these substances—rather than cups. Herodotus, the famous Greek historian, describes how the Scythians used hemp:

> On a framework of three sticks, meeting at the top, they stretch pieces of woolen cloth. . . . Inside this little tent they put a dish with red-hot stones in it. Then they take some hemp seed, creep into the tent, and throw the seed on to the hot stones. At once it begins to smoke, giving off a vapour unsurpassed by any vapor bath one could find in Greece. The Scythians enjoy it so much that they howl with pleasure.[9]

One archaeologist suggests that the original users of hemp were a group of farmers/herders in southern Russia, and that they "celebrated its importance by imprinting it on their pottery," which is impressed with cord designs, resembling hemp rope.[10] Similarly, some of the opium users of western Europe (the poppy was first domesticated in the western Mediterranean) used special containers shaped like poppy heads to hold their intoxicant of choice.

THE IMPLICATIONS OF THE SECONDARY PRODUCTS REVOLUTION

The spread of drinking and smoking cultures, along with the secondary products revolution, transformed life across Eurasia. Chapters 3 and 4 described some of the changes in state societies like Mesopotamia, Egypt, and the Indus Valley. We will now review what happened in peripheral societies—what changes the urban revolution inspired in Europe and in the central Asian plateau. Because central Asia has always served as the passageway between the Middle East and China, we will also look at the effects of the secondary products revolution on China in the third and second millennium BCE.

CHIEFLY SOCIETIES: THE PLOUGH AND PASTORALISM IN EUROPE

THE FIRST EUROPEAN FARMERS: Agriculture entered southeastern Europe (Greece and the Balkans) from southwest Asia by 6000 BCE and quickly spread into central Europe, until by 5500 BCE—a mere half-millennia later—farmers had colonized much of continental Europe. These early farming communities lived in a similar manner, whether in Germany, Hungary, or Serbia. Like farmers in southwest Asia, they used hoes to prepare the soil and then planted emmer wheat and barley. They chose to settle in very specific locations "where damp lowland pasture coincided with tracts of light, friable upland soils."[11] In such areas, a farmer could easily work the land with a hoe and produce a good harvest, year after year, while the family's cows could graze in the lowlands. Obviously, not many places meet these criteria—as a result, early farming communities were thinly scattered over much of the continent, separated from one another by thick forests. Still, European farming populations grew quickly; within 500 years, population densities in Germany rose from 1 person per 1,000 km^2 to 1 person per 120 km^2. These farmers lived in egalitarian communities where communal identities were emphasized.

HOES AND FEMALE VALUES: Anthropologists have established that women have a higher status in societies where the hoe, not the plough, is the main agricultural tool. Usually, women wield the hoe, tending the fields, and are therefore respected as "the breadwinners" of the family. This seems to have been true in early farming communities in central Europe. From 5500 to 4000 BCE, extended families—led perhaps by a matriarch—lived together with their domestic animals in longhouses built of timber. Archaeologists think that the different extended families that made up a hamlet or a village were also probably related, forming a lineage.

BURIAL MONUMENTS AND LONGHOUSES: By 4000 BCE, in northwest Europe (including Britain, France, Ireland, Denmark, and some of Germany and Poland) people began building monuments to house the dead. These tombs look surprisingly like the longhouses where people had been living for generations. They do not contain the remains of just one person; rather, they usually contain the bones of several people, often buried at different times. These tombs were probably elaborate family crypts, where the living met at certain times a year to commemorate the dead. In some places, the skeletons of many different people were placed inside or near the monument long after death, suggesting that burial rites took place only at certain times. Broken pots and incense burners, as well as different offerings, suggest that communal feasting and possibly opium use marked these funerals. Early farming societies emphasized equality and communality, as opposed to hierarchy and individualism.

THE PLOUGH AND THE SHIFT TO MALE VALUES: The plough gave early farmers a tool that allowed them to greatly expand the land that could be farmed. Along with ploughs, archaeologists find quantities of stone and later copper axes, which these farmers used to cut down the forests of Europe to turn them into fields. This ancient deforestation also provided excellent pasture for the sheep and goats that the secondary products revolution had suddenly made so useful. The new importance of domestic animals probably meant a rise in the status of men—who generally herded them. Because ploughs also require animal power, a shift from hoe agriculture to plough agriculture often means that men wind up in charge of farming as well. In Hungary, these innovations meant that two new cultures grew up side by side: farmers who continued to emphasize equality and the community and herding immigrants from the steppe. At a certain point, the herders introduced new customs to the settled societies, like respect for a warrior class. We can see this in the sudden appearance of individual burials of male warriors, which began among the herders and spread to the farming communities. People abandoned the tradition of communal tomb building, and women lost much of their status. Rituals and religions lost their meaning. Part of this may have had to do with the impact of new drugs, like alcohol. Instead of being incorporated into a community ritual, as opium had been before, certain powerful men seem to have enjoyed a drink with friends, or fellow warriors—perhaps before going off to pillage a nearby village. European potters began making special drinking vessels in the form of wagons or carts with oxen. Sometimes these fancy beer steins would find their way into warrior burials, along with metal weapons; this suggests that these European cultures linked the new agricultural regime, alcohol, and metallurgy. Andrew Sherratt believes that this shift in ritual is one from "female" values—the values of the communal, hoe-using cultures—to "male" values—the values of the new plough-based agriculturalists—from a focus on the home and the family to one on violence, animals, and alcohol.[12]

BARBARIANS AT THE GATE: NOMADS AND CHINA

NOMADS AND TRADE: In central Asia, the secondary products revolution allowed nomads to conquer the steppe. Because nomads regularly travel long distances and trade widely, they often introduce new goods and ideas from one place to another. In Eurasia, nomads are an unpredictable force that has facilitated contact between the west and the east. During the second and third millennia BCE, as sheep and goatherders moved into the mountains and deserts on China's northern and western boundaries, another, much slower, communication network emerged. These nomads, who doubled as traders, brought with them inventions from western Asia, which once modified in China, provided the symbolic basis for the first Chinese state during the Shang dynasty (1722–1121 BCE). Shang kings sought to control these new technologies, particularly wheeled vehicles and the horse, in order to cement their power over their subjects. Chinese artisans took bronze metallurgy—a Western invention—and gave it a Chinese imprint by inventing new techniques, which allowed them to make bronze vessels used in ancestor cults (see Chapter 4).

CHINESE ATTITUDES TOWARD NOMADS: Later written Chinese accounts show the distaste that the urban Chinese felt for their nomadic neighbors to the north and west. This loathing is apparent in a description of the Hsiung-nu, a famous group of powerful central Asian nomads who both raided and traded with the Chinese from around 300 BCE–200 CE.

> The Hsiung-nu live in the desert and grow up in a land which produces no food. [They are the people who] are abandoned by Heaven for being good-for-nothing. They have no houses to shelter themselves, and make no distinction between men and women. They take the entire wilderness as their villages and the *ch'inung lu* tents as their homes. They wear animal's skins, eat meat raw and drink blood. They wander to meet in order to exchange goods and stay in order to herd cattle.[13]

The Chinese may have believed that the nomads were primitive barbarians, unacquainted with houses, etiquette, fashionable silks, chopsticks—or anything else that made life worth living—but the nomads on the borders of China and "China" itself share a common point of origin—one which is intimately tied up with the secondary products revolution.[14]

EMERGENCE OF PASTORALISM AND METALLURGY IN CENTRAL ASIA: Before about 3000 BCE, the inhabitants of the steppes and deserts of central Asia were hunter-gatherers. The hostile landscape with almost no water and extreme temperature variation did not lend itself to farming. The wide plains of central Asia could provide good grazing for herds of animals, but before the spread of wheeled transport, milking, and the horse, becoming full-time nomads was nearly impossible—or at least not a desirable alternative to a foraging lifestyle. The secondary products revolution provided pastoralists with the tools they needed to prosper on the central Asian steppe. At the same time, in the early third millennium BCE, copper and bronze production began in central Asia, perhaps in response to developments in the emerging urban centers on the periphery of this area: places like Mesopotamia, the Indus Valley, and even along the southern fringes of central Asia, in the highlands of southern Turkmenistan. Generally, communities that specialized in mining copper and tin were located hundreds or thousands of kilometers from communities of bronze smiths. The emergence of metallurgy allowed pastoralists to become middlemen in the bronze trade. By the early second millennium BCE, these herders began to move to mountain passes, like Djungaria and Semerechiye in Kazakhstan, where they spent the summers in the cool mountain meadows and the winters with their herds on the surrounding plains. Nomads supplemented pastoralism with a little pocket money earned by taxing traders traveling from the mining regions in the mountains to the copper production centers in the plains below. As Michael Frachetti explains, "specialized herd management in the foothill zones of the eastern steppe developed in tandem with negotiations of trade and the political control of the regional corridors that facilitated the transfer of human, animal and material resources."[15]

TARIM BASIN COMMUNITIES: At roughly the same time that pastoralists began to control metal sources in Kazakhstan, other pastoralists/farmers moved into the Tarim basin and established a few oasis towns. The mummies of some of these people, naturally preserved by the salty, dry sand of the Taklimakan desert, amazed Western archaeologists when they were discovered in the 1980s (and later reported to the world) because they looked non-Chinese. Instead, they had blonde, red, and light-brown hair. They probably spoke an Indo-European language, Tocharian, a language related to English, Spanish, Persian, and Hindus that was probably spoken only in eastern China. Very few excavations have been undertaken in western China, Siberia, or central Asia, making it difficult to accurately trace the origins of the first inhabitants of the Tarim basin.

The sedentary inhabitants of the Tarim Basin in eastern central Asia interacted with people in inland China and introduced some Western practices to China, and some Chinese practices to the West. Essential imports to China include basic nomad technology—like fast horses and carts—as well as the new central Asian specialty: bronze. Chinese smiths quickly learned how to make bronze themselves, inventing and perfecting new techniques to make both weapons and elaborate vases, but the first bronze seems to have entered by way of the Tarim basin.[16] Oasis communities in the Tarim basin were cosmopolitan, building their houses out of mudbricks, like people in western Asia, central Asia, and India, but surrounding them with walls made out of stamped earth, a Chinese technique. They grew both barley (a west Asian crop) and millet (a Chinese domesticate), while some enterprising immigrants made wine for the Chinese export market using grapes imported from Iran. Ethnic Chinese and different central Asian groups lived together, intermarrying, and communicating in a dizzying array of languages.

THE JADE ROAD AND CHINESE CONTACTS WITH THE WEST: The northern and western pastoralists also traded with the Chinese. Thousands of years before the Silk Road, the Chinese opened up the "Jade Road" to procure jade from east central Asia, a trade that pastoralists probably oversaw in much the same way as the Kazakhstan copper trade. Some of these people, shamans, who used ephedra, opium, or hemp for their trance-inducing powers, seem to have traveled to China as mages, a Persian word also found in English, which became the ancient Chinese word for magician as well—mag. Carvings of these sorcerers have been found at the Shang and Zhou courts, suggesting that the Chinese made use of both practical and esoteric Western inventions.

Yet, contact between the nomads and the Chinese was not always mutually beneficial. Nomads have a complex relationship with farmers and state officials. Because they spend most of their lives outside of state control, they can represent both an alternative to state control and a threat to state armies. At the same time, nomads need villages to survive. Nomads do not just live off of milk and meat; in fact, grain is an essential part of their diet.[17] If villagers let nomads trade their animal products for agricultural products, the two groups can live peacefully. But if something happens to the frontier markets, nomads will raid villages to obtain the same products. Whether nomads trade or raid also depends on the balance of powers between them and the state. When nomads are strong, and the state is weak, they will raid, and sometimes even conquer states. This uneasy symbiosis between nomads and towns in China is the main reason that China built the great wall.

In the thirteenth century CE, Chinggis Khan and his descendants created the largest continuous land empire ever. It spread from Indonesia to Hungary and encompassed the two most powerful Asian states of the time, China and Persia. Chinggis Khan is only one descendent of a long line of powerful nomadic tribes who have always played a major role in Eurasian history. Because the steppe of central Asia bordered nearly every Eurasian civilization in antiquity, nomads have been critical to the spread of ideas and goods, often founding and controlling trade routes.

AMERICAN NETWORKS

A TRANS-AMERICAN WORLD SYSTEM?

If the secondary products revolution spread so quickly across Europe and Asia, leaving chiefdoms and states in its wake, what about North and South America? Did they experience an equivalent revolution at any point? Did innovations like gold metallurgy, intensive corn production, or tobacco smoking ever spread across these continents stimulating social change? In short, was there a network of relationships in the Americas similar to that in the Afro-Eurasian world? There are no simple answers to these questions.

SLOW NORTH-SOUTH DIFFUSION: First, unlike in the Old World, geographic factors made the spread of agricultural technology difficult. The major axis of the Americas is north-south, whereas for Eurasia it is east-west. This means wide areas of Eurasia lie on the same latitude and experience similar growing seasons and climates. Barley can be easily grown from Egypt to Mesopotamia to India. In the Americas, on the other hand, ancient civilizations lay on different latitudes with vastly different seasons (just consider the differences between the Andes, Mexico, and the northeastern United States). In order to grow maize outside of the tropics, it is necessary to genetically engineer strains that are adapted to the short growing season of southern Canada, for example. This was hard to do. Genetic analysis of American crops, like the lima bean, common bean, chili pepper, and squash, shows that they were all domesticated more than once in different parts of South and North America. This shows how slowly crops spread, because it is much easier to grow an already domesticated crop than to start from scratch. In contrast, all Eurasian crops seem to have been domesticated only once. This north-south axis also affected the diffusion of other inventions. Writing, invented in Mesoamerica, never in 2000 years spread to the Andes, and the quipu system never spread

The American World System—a map showing the Americas, with Cahokia, Snaketown, Chaco Canyon, and Tollan all marked, along with arrows illustrating the spread of maize from Mesoamerica, Mississippian trade routes, and Pochteca trade routes.

north. In Eurasia, on the other hand, the invention of writing in Mesopotamia was followed by its almost simultaneous appearance elsewhere in the world: in Egypt, Syria, western Iran, eastern Iran, Turkmenistan, and the Indus Valley. Contact was so minimal that in 1500 CE, the Aztec Empire and Tinhuantinsuyu were basically unaware of each other's existence.

RELUCTANCE TO RESEARCH CONTACT: Second, even when there is evidence for contact, archaeologists may hesitate to research it. Archaeologists today are reacting against the early history of archaeology in North America, when scholars were willing to believe that *anyone* other than the ancestors of the Native Americans had been responsible for building the mounds, carving the figurines, or polishing the beads that they found in excavation. As a result, scholars may reject even the idea of "contact" because they dislike its racist implications. Even in cases where there is clear evidence for the spread of innovations, for example, the spread of maize to the southeastern United States, its significance is downplayed. We can see how this works by looking at Cahokia, an ancient town in Illinois, and evidence of its trade connections to Mexico.

THE CORN GODDESS AND CAHOKIA

EARLY REACTIONS TO CAHOKIA: Across the Mississippi River from St. Louis lies the largest archaeological site in the present-day United States. The mounds that make up Cahokia are spread over an area of 13–16 square kilometers—making this settlement almost as large as Teotihuacan in Mexico and 15 times larger than classical Athens. In fact, Cahokia was so large that it took archaeologists a long time to recognize it as one ancient town. They simply could not believe in its dimensions. In 1811, less than 10 years after Thomas Jefferson acquired this land as part of the Louisiana Purchase, one of his friends, Henry Brackenridge, traveled to St. Louis and surveyed the ruins of Cahokia. In a letter to Thomas Jefferson, he describes these ruins:

> What a stupendous pile of earth. . . . When I arrived at the foot of the principal mound I was struck with a degree of astonishment, not unlike that which is experienced in contemplating the Egyptian pyramids. To heap up such a mass must have required years and the labor of thousands. . . . If the city of Philadelphia and its environs were deserted, there could not be more numerous traces of human existence."[18]

Throughout the nineteenth century, historians and archaeologists argued about the identity of the people who built Cahokia and other mounds scattered throughout North America. Scholarly consensus held that a race of moundbuilders—who had long disappeared from North America—had constructed these giant mounds, rather than Native Americans. J. W. Foster, president of the Chicago Academy of Sciences, summed up the general opinion when he wrote that assuming that the Indians were responsible "is as preposterous, almost, as to suppose that they built the pyramids of Egypt."[19]

SIMILARITIES BETWEEN CAHOKIA AND MESOAMERICA: Even when most scholars had quietly accepted that Native Americans had built the mounds, many of them still thought that Cahokia emerged because of outside influence from Mesoamerica. From 1050–1250 CE, when Cahokia was at its height, a temple or palace probably stood atop Monk's Mound—the largest earthen platform in North America. South of this mound lay the great plaza with other mounds, which probably served as the burial places of important families. A wooden palisade enclosed this ceremonial area, often referred to as "downtown Cahokia." Outside of this palisade lay several distinct neighborhoods: wooden houses grouped around mounds and plazas. Each neighborhood probably housed people who either belonged to the same lineage or practiced the same profession. The inhabitants of the northernmost suburb, for example, specialized in drilling shell beads and polishing stone celts. The arrangement of plazas and mounds in downtown Cahokia reminded many people of the layout of Teotihuacan, with its elaborate stone pyramids and plazas, and the special neighborhoods of craftworkers also seemed reminiscent of Mesoamerica.

Mississippian Agriculture and Economy: Moreover, the people who lived in Cahokia, and other communities along the flood plain of the Mississippi and its tributaries, relied on intensive cultivation of maize, squash, and beans—the Mesoamerican agricultural trinity. These farmers built raised furrows to protect the crops from flooding and to assure bountiful harvests. Although the Mississippian people were not the first farmers in the eastern United States, the cultivation of corn allowed them to adopt a completely sedentary lifestyle for the first time. Farmers in the eastern United States had begun to domesticate native plants by 2200 BCE, but none of the available native plants—sunflowers, sumpweed, goosefoot, and knotweed—allowed them to give up a foraging lifestyle and build sedentary villages. The Cahokians did continue to grow many early domesticates and ate a diet that included fish, game, and some seasonal wild plants, but corn cultivation allowed them to amass large surpluses of food, which they stored in newly designed communal granaries. Although the first corn arrived in the eastern United States by 200 CE, it remained an unimportant crop until approximately 900–1000 CE—when a new variety, suited to the short, humid summers of the Mississippi River appeared. We can see the fundamental importance of corn to people who had been used to tending sumpweed—a smelly, hay-fever-inducing relative of ragweed—in the stories of some of the tribes that once lived near Cahokia, which emphasize the importance of the Corn Mother.[20]

Shared Corn Rituals: One of the most striking things about corn myths and rituals from the American southeast is their similarity to other such traditions in Mesoamerica. The Cherokee word for both corn and the corn goddess is "Selu", which probably comes from the Nahuatl word "xilo." Nahuatl was the language spoken in central Mexico by the Aztecs, and perhaps the Toltecs. In both Puebla, Mexico, and among the Mississippians, the green corn ceremony was the most important religious event of the year. New corn could be eaten only after the completion of the ceremony, which took place just as the ears began to ripen. In Puebla, a priest slit the throat of a young girl who represented the corn goddess, Xilonen, as she lay on a platform made from the bodies of four young men, whose hearts had been removed after they were placed on glowing coals. At Cahokia, a set of graves containing four men whose hands and heads had been removed may have formed a similar platform; one of the other bodies in the same pit may have been a young girl representing Selu.

SOUTHEASTERN CEREMONIAL COMPLEX: Did corn arrive in Cahokia from Mexico as part of a larger belief system? Did the traders who brought these multicolored ears also bring the practical and ritual knowledge of how to cultivate it? Archaeologists refer to a set of interlocking beliefs—expressed in burial practices, art, and settlement placement—as the Southeastern Ceremonial Complex. Elements of this "cult" sometimes seem to echo Mesoamerican practices. A figurine found in Cahokia, for example, depicts an agricultural goddess "hoeing the back of a jaguar muzzled serpent whose tail dissolved into gourds."[21] Jaguars are a bit out of place in Illinois, but they fit in perfectly with traditional Mesoamerican beliefs. Some scholars have even argued that certain cultic objects with Mesoamerican themes "were brought to Mississippian country by professional Mexican merchants, the famous *pochteca*."[22] The pochtecas were a guild of Toltec traders who plied their wares and gathered intelligence in distant communities, far from the capital of Tula, north of present-day Mexico city. We suspect that they traveled as far north as northern New Mexico to trade macaws and copper for turquoise and as far south as Guatemala to trade obsidian for glazed pottery.

ARGUMENTS AGAINST MESOAMERICAN INFLUENCE: So is Cahokia somehow the result of Mesoamerican intervention? No one believes that Cahokia was built and settled by colonists from the valley of Mexico, but some people have argued that some of the ideas behind Cahokia—ideas of urbanism, of kingship, and of hierarchy—had their roots ultimately in Mesoamerican experiences. In the 1940s, a group of American archaeologists set out to excavate some earthen mounds in northern Mexico to see if they could trace the Mexican ancestors of this Mississippian tradition. Although they found some parallels, similarly decorated pottery and

FIGURE 8

The Birger figurine, found near Cahokia, is a bauxite statue showing an agricultural goddess hoeing the back of a jaguar-headed serpent whose tail dissolves into a squash-vine.

pipes, the vast desert that separated Mexico from Texas, and Texas from the easternmost Mississippian group made them reject any notion of Mesoamerican ancestry for Cahokia. Many of the parallels between Mesoamerica and Cahokia are shared by cultures around the world. Take mound building, for example—people all over the world seem to get a kick out of piling up earth or stone. There are pyramids in Egypt and Mexico, earthen platforms in England and Ohio, and mudbrick platforms in Syria and India. Similarly, most towns and cities have had a quarter devoted to particular artisans; this is as true of Mesopotamia as it is of both the Mississippi cultures and Mesoamerica. Additionally, there is no direct evidence of Mesoamerican artifacts or trade goods in Mississippian communities. There is no obsidian from central Mexico, or Toltec pottery, or macaw skeletons. Similarly, there are no specific Mississippian products in Mesoamerica. Even if some pochtecas did make the 3,000 mile trek to Cahokia, they cannot have done so often enough—or stayed long enough—to have much of an impact on Mississippian life. As a result, in the last 50 years, most archaeologists working in the southeastern United States have rejected out of hand any arguments for Mesoamerican contact with Cahokia.

CORN: FROM SNAKETOWN TO CAHOKIA: So what about the corn rituals and stories? What about the corn, squash, and beans itself? Or the corn goddess figurine at Cahokia? Corn probably came from an intermediate area rather than entering as a direct import from Mesoamerica. Although corn originally entered what is now the United States in about 1000 BCE, sedentary farming villages based on maize cultivation did not appear until 1 CE. The first maize farmers lived in small villages in the Sonora desert of southern Arizona and used vast irrigation systems to water their crops. After about 600 CE, these farming villages began to grow much larger, and clear and continuing evidence of contact with Mexico emerged. At Snaketown, near Phoenix, Arizona, residents constructed a large ball court and imported rubber balls to play the popular Mesoamerica ballgame. Casas Grande, a prehistoric town in Northern Chihuahua, Mexico, may have served as the point of contact between Mesoamerican traders and local southwestern groups. Casas Grande belonged to the same tradition as Snaketown, but had close ties to a settlement that housed immigrants or traders from the Valley of Mexico. Further to the north, farmers living at Chaco Canyon and other Anasazi communities in New Mexico and southern Colorado from 700–1200 CE also grew maize. This area, called the Four Corners Region, is a high plateau between 1,500 and 2,000 meters above sea level (5,000–7,000 feet above sea level). As a result, summers, although hot, can be quite short. To successfully farm maize, these communities had to develop strains that could withstand short summers and the extreme temperature variations of this plateau. Mississippian societies probably encountered these strains, which could also be grown during the short, humid summers in the American bottoms through one of the exchange networks that loosely connected groups throughout the present-day United States. Descriptions of Indian trading depots along the Mississippi after European contact show how such a system may have worked:

> Originally a market where corn, beans, and squash were exchanged for meat and hides, the farming villages on the upper Missouri developed into an exchange center for European goods and Plains produce as well and for guns and horses in particular. . . . Assiniboines, Crees, Ojibwes, Crows, Blackfeet, Flatheads, Nez Perces, Shoshones, Cheyennes, Arapahos, Kiowas, Kowas Apaches, Pawnees, Poncas, and various Sioux bands all visited the upper Missouri villages, either regularly or intermittently. These visitors passed on what they obtained to more distant neighbors, often at vastly inflated rates. In this way the Mandan-Hidatsa-Arikara trade center interconnected with other

exchange centers—the Wichita, Caddo, and Pawnee on the prairies to the south; the Comanche network on the southwestern plains; the Pueblos on the Rio Grande; the Columbia River networks beyond the Rocky Mountains; and via British traders and Montreal, the fur houses and markets of Europe."[23]

This scenario relies upon "down-the-line" trade—in which communities may receive a certain item, or learn of a certain innovation from their neighbors, which they then pass onto other neighbors. Over time, goods and ideas can travel long distances, according to the same scenario that we reconstructed to show how luxury goods and the secondary products revolution diffused throughout Eurasia.

NEW EVIDENCE FOR CONTACT BETWEEN CAHOKIA AND MESOAMERICA: But new evidence has forced certain archaeologists to reconsider Cahokia's relationship with communities in Mexico. Rather than suggesting Mexican influence on Cahokia, however, they suggest that Cahokia may have influenced communities as far south as northern Mexico. The Huasteca site of Tantoc contains more than a dozen earthen mounds, whose orientation, both to each other and to other architectural elements, strikingly parallels certain Mississippian communities. Moreover, the artifacts found at Tantoc have no parallels to those found elsewhere in Mexico, but they are strikingly similar to material that was used by the westernmost Mississippian groups in eastern Texas. Tantoc is dated to slightly after Cahokia's rise. It seems possible that merchants from these Texan groups traveled along the barrier islands of the Gulf coast and down the Tampico River to Tantoc. Instead of Toltec pochtecas bringing their ideas of civilization along with macaw feathers or red stone figurines to the Mississippi, the Mississippians may have exported some of their "foreign" ideas to Mexico.

The example of Cahokia should caution us about oversimplifying "contact." The centuries between roughly 900 and 1250 CE witnessed the rise and fall of Cahokia along the Mississippi, Chaco Canyon in New Mexico, and the Toltec state in Mesoamerica. The connections between these three cultures are unclear, although it seems certain that no one culture was instrumental in the rise of any others. Each one developed from an indigenous base, although contacts between all three may have stimulated aspects of the development of each. Melvin Fowler, a prominent archaeologist who specializes in Cahokia, believes that "to fully understand Cahokia the entire complex of Pre-Columbian civilizations, like pieces of a hemispheric three-dimensional puzzle, must be judiciously fitted together."[24] At the same time, researchers must keep an open mind and realize that contact and influence never operated along just one direction. The evidence from Tantoc means that we must be careful not to assume that "more complex" civilizations—like the Mesoamericans with their traditions of states and writing—automatically influenced "less complex" civilizations like Cahokia.

THE COLLAPSE OF AN ANCIENT NETWORK

MEDINET HABU AND "THE CATASTROPHE":   If we can trace the emergence of these ancient world systems, we can also study their dissolution. In the Egyptian desert, not far from the Valley of the Kings, stands one of the mortuary temples of the pharaoh Ramesses III. Preserved by the sands of the Sahara, the temple at Medinet Habu looks much as it did when it was built in approximately 1170 BCE. The reliefs and accompanying inscriptions on this miraculously preserved building tell the story of the end of a world:

> . . . As for the foreign countries, they made a conspiracy in their sealands. All at once the lands were on the move, scattered in war. No country could stand before their arms. Hatti, Kode, Carchemish, Arzawa, and Alashiya. They were cut off. A camp was set up in one place in Amor. They desolated its people, and its land was like that which has never come into being. They were advancing on Egypt while the flame was prepared before them. . . ."[25]

This is the obituary to an ancient world. In the early years of the twelfth century BCE, the great Bronze Age civilizations of the Mediterranean and western Asia ended. Cities on the Greek mainland (Mycenae), the Greek Islands, Anatolia, and the Levant were destroyed in violent conflagrations. Their populations fled the destruction and never returned. Historians refer to this period as "the crisis years," or "the catastrophe." In Greece, Anatolia, and Syria, not only cities, but also the trappings of civilization disappeared. Writing was forgotten for hundreds of years as people abandoned cities to settle in small villages. The entire palace system, which supported hundreds of diverse craftsmen, scientists, and bureaucrats, and employed and fed thousands of farmers, disappeared. Even the few great civilizations that survived—Assyria, Babylonia, Elam (in Iran), and Egypt—were affected by this catastrophe. Their scribes grew silent, as between about 1200 and 900 BCE, a dark age spread over the world of states. When written records appear again, history had forgotten about many of the Bronze Age civilizations. They were rediscovered only in the nineteenth century as archaeologists began to excavate ancient ruins. Prior to their rediscovery, only a few garbled stories attesting to both the achievements of these now mythical civilizations, and of the catastrophe itself, survived. Two of these stories remain at the heart of "Western civilization," the Homeric epics and the story of Exodus. So much was lost that it is easy to agree with Robert Drews when he states that "altogether the end of the Bronze Age was arguably the worst disaster in ancient history, even more calamitous than the collapse of the western Roman Empire."[26]

Who burnt these cities, and why? Why was the destruction so widespread? What factors made the collapse of these kingdoms, and the entire world system they represented, permanent? Before we can answer these questions, we must understand both the Bronze Age world system that the catastrophe destroyed, a set of civilizations that we have already described in Chapters 3 and 4, as well as the Iron Age civilizations that appeared after the dark age. People in Israel, Greece, and Rome all sought their origins in the story of the catastrophe. Their civilizations, with the use of "alphabetic writing systems . . . , of republican political forms, of monotheism, and eventually of rationalism" often serve as the starting point for Western civilization.[27]

AMARNA DIPLOMACY

DIPLOMACY IN THE MEDITERRANEAN, 1500–1200 BCE: Before the catastrophe, Mediterranean civilization had an international system that regulated relations between states. Like our own international law, states in the late second millennium BCE relied on standard practices to avert conflict and ensure peace. Tablets discovered at Akhenaton's capital, Amarna, in Egypt, preserve the international correspondence of the fourteenth century BCE. Akhenaton corresponded with the kings of a range of different states: great powers like Babylonia, Hatti (in present-day Turkey), and Assyria, smaller states like Alashiya (Cyprus), and his vassals along the Levantine coast. The correspondence underlines that the great powers did everything possible to avoid fighting each other openly, as such battles could have been devastating to both parties. Instead, they sent each other royal gifts, contracted marriage alliances, and wrote letters affirming their good intentions.

HATTI AND EGYPT: Relations between Hatti and Egypt in the century before Hatti's destruction illustrate how this world order worked. The culminating battle of the "world war" of the thirteenth century BCE took place at Kadesh in Syria in 1285/6 BCE between these two states. Both Egypt and Hatti were growing empires that found the rich coastal cities irresistible. From 1500–1300 BCE, Egypt controlled most of the southern Levant—what is now Israel and Palestine—but Hatti was slowly extending its tentacles south, bribing Egypt's vassals to switch their alliances. Although Ramesses II claims to have won an overwhelming victory at Kadesh, we can discount his bluster, as the Hittites were left in control of most of the Levant following a treaty made between the two powers. After this battle, Egypt and Hatti never fought each other directly again, but continued to stage proxy wars between their vassals in the Levant. Smart Levantine kings played Hatti and Egypt off each other—sometimes accepting the visitors of one but not the other—in much the same way that various "non-aligned" countries in Africa, the Middle East, and Asia manipulated the U.S. and the Soviet Union during the cold war.

In the Mediterranean, the century prior to the catastrophe was one of peace. Although wars were fought, they were not fought between the great powers, who traded peacefully and maintained a range of diplomatic contacts. The Myceaneans exported wine and olive oil to the eastern Mediterranean, while Cyprus supplied copper (for bronze) and Iran contributed tin. In the cities, art and literature flourished, and there was a shared international language, Akkadian, in which various popular myths and stories circulated throughout the ancient world.

MARAUDERS AND MERCENARIES: Just outside of the main purview of the states, but still within the Bronze Age world, were the "barbarians" living in the Balkan, Ukraine, and the western Mediterranean. All of the great powers were forced to fight these marauders occasionally, usually at their borders. Year after year, the Egyptian army defended the delta against the Libyans to the west, while the Hittites constantly had to fight the Kaska tribes to the north. The great powers also, however, employed mercenaries who were drawn from these barbarian tribes. The Egyptians and the Hittites, for example, both recruited the "Sherdan" infantryman—whose name indicates that they originally came from Sardinia.

THE CATASTROPHE

During a 10-year period, from about 1190–1179 BCE, at least 47 cities were destroyed. The shadowy Sea Peoples of the Medinet Habu inscriptions had entered the Mediterranean and begun their destruction. How did they succeed? The maintenance of the diplomatic system previously described relied upon wide contacts among different groups: empires, vassal states, and "barbarians" hailing from chiefdoms and simple villages. Yet the interactions that produced the Bronze Age diplomatic system may have also contained the seeds of its destruction. One theory blames the catastrophic collapse of these Mediterranean civilizations on barbarian mercenaries employed by the great powers. Around 1200 BCE, these mercenaries, well acquainted with the weaknesses of the large state militaries, turned the tables. Using tactics that we might refer to as "guerilla warfare" or "terrorism" today, they swiftly brought down these state armies, winning battle after battle and destroying the previous world order.

MILITARY INNOVATIONS AND GUERILLA WARFARE: To understand the origins of the Sea Peoples, scholars have considered a wide range of evidence—the inscriptions in Egypt, bas-reliefs depicting the battles, and letters written just prior to the destruction of cities in the Levant. After carefully weighing the various theories proposed for the end of the Bronze Age, Robert Drews argues that innovations in military technology led to the catastrophe. During most of the Late Bronze Age, armies of the great powers relied on chariot warfare to defend their cities from barbarians and to fight other great powers. The chariot had been a major technical innovation when it was introduced a few hundred years before, because it allowed archers a mobile platform from which to shoot their opponents. A small number of well-trained archers and charioteers, drawn from the nobility of the great powers, fought most wars. Infantry had almost no role in them, except to guard the camps and horses during the night from sneak attacks and to protect stationary chariots during battle. The only infantrymen who participated in the battles were soldiers who ran along the chariots and engaged in the hand-to-hand fighting that ensued after the chariots were derailed. These runners were generally mercenaries—hired from the barbarian lands—as these men had more experience in battle than the inhabitants of large Near Eastern cities or peasants from the peaceful countryside.

The catastrophe occurred when barbarians, many of whom probably had been trained as mercenaries by the great powers, realized that they could overthrow a force of chariots by using infantry. Because chariots were fragile and always in danger of toppling over, they moved slowly. A runner armed with a javelin could strike a chariot more easily than it could strike him. The barbarians imported great quantities of a new type of sword developed in the Balkans, one that could be used both to cut and to thrust. These swords were broader and more substantially built than the rapiers that had previously been used in hand-to-hand combat in the Middle East. With the new sword, warriors could quickly cut off an opponent's head, leg, or arm, or cut him in two; the old swords were not as effective. Egyptian reliefs depicting battles with the "Sea Peoples" show a new emphasis on infantry, due to a growing realization that chariots were no longer an effective way to counter these new armies. Joshua 11.1-11 also preserves an account of the final days of Hazor, a city that was destroyed during the catastrophe by barbarian infantrymen, in this case Israelites, attacking the chariots of this Levantine kingdom. The Illiad also tells the story of the victory of a barbarian army relying on infantry, the Greeks, against the urban Trojans and their horse-reliant allies.

In the end, the kingdoms fell to mercenaries that they themselves had trained and armed. These barbarians won, not by using the expensive technology of the states, like chariots, but by using low-tech weapons—just men armed with javelins and swords. They destroyed the cities by making their inhabitants realize that large wealthy centers were vulnerable; uneducated warriors could best the professional armies of the kingdoms. The great kingdoms of Mycenaean Greece, Hatti, and the rich cities of the Levant collapsed because the rules of engagement had changed profoundly in a way they never suspected. A letter written by the last king of Ugarit (a city on the Syrian coast) to the king of Alashiya (Cyprus) illustrates the vulnerability of the professional armies of these kingdoms faced with attacks from the Sea Peoples. This city, which was a vassal-state of Hatti, fell because its professional army was off fighting the barbarians in Anatolia when the Sea Peoples attacked:

> To the King of Alashiya, my father, from the king of Ugarit, his son. I fall at my father's feet. . . My father, the enemy ships are already here, they have set fire to my towns and have greatly harmed the country. Did you know that all my troops were stationed in Lycia and have not yet returned? So the country is abandoned. . . Consider this, my father, there are several enemy ships that have come and done great damage. Now if there are more enemy ships let me know about them, so that I can know the worst.[28]

According to Ugarit's excavator, this letter was found, still waiting to be posted, in the ruins of the palace at Ugarit, which had been destroyed before aid could reach it.[29]

DROUGHT AND "THE CATASTROPHE": Why did the Sea Peoples begin their crusade against the east Mediterranean states? And why were the kingdoms never re-established once the threat posed by these invaders had subsided? Lake sediments in Turkey and millennia-old ice in Greenland both indicate that a little Ice Age began at about 1200 BCE, which would have led to cold rainy weather in northern Europe and drought in the Mediterranean and eastern Europe. This would explain both why the marauders attacked the cities in the first place, because drought made them leave their original homes, and why the kingdoms were never rebuilt. If there was a widespread drought, the palaces would not be able to reap the agricultural surpluses that were necessary for their existence. Agriculture would have devolved into subsistence farming. This also explains why the dark ages were so general, and affected areas that were not destroyed in the catastrophe—like Mesopotamia and Egypt.

SMALL POX AND THE END OF THE BRONZE AGE: Another possibility is that a catastrophic epidemic, or a series of epidemics, ravaged Bronze Age communities in the eastern Mediterranean (and perhaps as far east as China). There is little historical or archaeological evidence to support this theory, beyond vague references to plagues, with one exception. Ramesses III's son, Ramesses V, died of smallpox; the characteristic pock marks still mar his mummy's face. Smallpox, a highly lethal and contagious disease, represented a major threat until its eradication in the twentieth century. Epidemics of smallpox may have been behind some of the worst epidemics in the ancient world, like the plague that hit Athens in 430 BCE, the Antonine plague in the Roman Empire in 165–180 BCE, and the plague of Cyprian in 251–266 BCE. Some passages in the Hebrew Bible may also refer to smallpox. Numbers 14 relates that the Israelites who visit Canaan on a scouting mission contract a plague, which quickly kills all but two men over the age of 20. Elsewhere in Exodus and Numbers, people fall victim to a disease that is described as a swelling or rash—perhaps smallpox. The evidence for a smallpox epidemic—or a series of epidemics during the twelfth century is slight, but tantalizing. Like the Black Death during the fourteenth century CE, smallpox could have stowed away on trading ships or in donkey caravans. It may have spread along trade routes, and so the intense diplomacy and trade of the Amarna World System could have helped it spread across western Asia, northern Africa and Europe.

THE RISE OF THE POLIS SYSTEM: The end of the Bronze Age introduced profound changes across the Mediterranean. In the Iron Age societies that arose centuries later, warfare became everyone's concern, and infantry became the main component of all ancient armies. The new cities were usually highly defended fortresses that could withstand sieges better than interior cities. Trade networks disintegrated, leading to the adoption of iron tools, as copper and tin supplies were no longer easily available. When civilization revived after the dark ages, "the kingdom system" disappeared in Greece, to be replaced by a "polis system."

CONCLUSION

The origins of the state in Eurasia, and, to a lesser extent, in Mesoamerica, had widespread consequences. The emergence of cities changed trade and economic relationships across Eurasia. The diffusion of the secondary products revolution—the plough, wool, milk, and metallurgy—had different consequences in different parts of the world. In Europe, the spread of the plough and sheep pastoralism led people to leave their ancestral villages and colonize new areas. Like pioneers settling the American West 5,000 years later, these farmers set out to subdue a continent, cutting down its forests, and probably battling its hunter-gatherer inhabitants. In central Asia, the secondary products revolution led to the creation of a nomadic way of life; hunter-gatherers became the cowboys of the east, who specialized in part-time metallurgy. In China, the introduction of the horse and the chariot did not have much practical effect. The inhabitants continued growing millet and rice and did not use pastoral products, but it did have a symbolic effect, which allowed a new ruling class to arise. The effects of the urban revolution are not as clear in Mesoamerica, where geographic factors prevented the easy spread of inventions. Still, at about 900 CE, the simultaneous appearance of three complex civilizations, all of which shared common traits, means that it might be time to look at events in North America from a macro-perspective as well. Of course, the establishment of a world system did not just rely on trading relations. By 1200 BCE, the states located around the Mediterranean had complicated diplomatic ties, rendering them interdependent, and also creating a network which encompassed people from nearby chiefdoms. When the crisis developed, each of these states was vulnerable to collapse precisely because of these close relationships. These case studies illustrate the long history of ancient worlds, and the long shared history of many civilizations.

TIMELINE

	5,000 BCE	4,000 BCE	3,000 BCE	2,000 BCE	1,000 BCE	0 CE	1,000 CE	2,000 CE
West Asia		Invention of the plough Invention of the wheel?	First evidence for milking First evidence for wool production Invention of alcohol Rise of Mesopotamia		International diplomacy The catastrophe			
East Asia				The spread of the horse and metallurgy to China	State in China Tarim Basin mummies Jade Road			
South Asia				Indus Valley civilization				
Central Asia		Invention of the wheel?	First domestication of the horse Domestication of hemp	Copper and bronze production begins in Central Asia	Jade Road			
Southeast Asia								
Australia								
Europe		Invention of the wheel? Invention of metallurgy	Alpine Ice Man First use of the plough in Europe Domestication of the poppy Shift from female values to male values	Wool enters Europe	The catastrophe	Rise of the Polis		
North Africa			First domestication of the donkey Rise of Egypt		International diplomacy The catastrophe			
West Africa								
East Africa								
South Africa								
North America				Domestication of sumpweed, sunflower, goosefoot and knotweed	Maize enters Southwestern US		First Southwest villages based on maize Snaketown Chaco Canyon Maize cultivation begins on Mississippi	Snaketown Chaco Canyon Casas Grande Cahokia
Meso-america								Toltecs
South America								

NOTES

CHAPTER ONE

1. M. H. Moncel, "The Discovery of the First Hominids in East Africa," *Anthropologie* 107, no. 2 (2003): 309–310; M. G. Leakey, F. Spoor, F. Brown, P. N. Gathogo, C. Kiarie, L. N. Leakey et al. "New Hominin Genus from Eastern Africa Shows Diverse Middle Pliocene Lineages," *Nature* 410 (2001): 433–40; Y. Haile-Selassie, "Late Miocene Hominids from the Middle Awash, Ethiopia," *Nature* 412, no. 6843 (2001): 178–181; G. WoldeGabriel, Y. Haile-Selassie, P. R. Renne, W. K. Hart, S. H. Ambrose, B. Asfaw, G. Heiken and T. White, "Geology and Palaeontology of the Late Miocene Middle Awash Valley, Afar Rift, Ethiopia," *Nature* 412, no. 6843 (2001): 175–178; B. Senut, M. Pickford, D. Gammercy, P. Mein, K. Cheboi and Y. Coppens, "First Hominid from the Miocene (Lukeino Formation, Kenya)"; "Premier hominidé du Miocène (formation de Lukeino, Kenya)," *Comptes Rendus de l'Académie des Sciences— Series IIA—Earth and Planetary Science* 332, no. 2 (2001): 137–144; M. Pickford and B. Senut, "The Geological and Faunal Context of Late Miocene Hominid Remains from Lukeino, Kenya"; "Contexte géologique et faunique des restes d'hominidés du Miocène supérieur de Lukeino, Kenya," *Comptes Rendus de l'Académie des Sciences—Series IIA—Earth and Planetary Science* 332, no. 2 (2001): 145–152.

2. For the discovery of *Homo floresiensis* see: P. Brown, T. Sutikna, M. J. Morwood, R. P. Soejono, Jatmiko, E. W. Saptomo and R. A. Due, "A New Small-Bodied Hominin from the Late Pleistocene of Flores, Indonesia," *Nature* 431, no. 7012 (2004): 1055–1061; M. J. Morwood, R. P. Soejono, R. G. Roberts, T. Sutikna, C. S. M. Turney, K. E. Westaway, W. J. Rink et al., "Archaeology and Age of a New Hominin from Flores in Eastern Indonesia," *Nature* 431, no. 7012 (2004): 1087–1091. For the debate over this fossil see: M. Kohn, "The Little Troublemaker" *New Scientist* 186, no. 2504 (2005): 41–45.

3. I am indebted to Andrew Sherratt for this notion. A. Sherratt, "Plate Tectonics and Imaginary Prehistories: Structure and Contingency in Agricultural Origins," *The Origins and Spread of Agriculture and Pastoralism in Eurasia* (London: UCL Press, 1996): 130–140.

4. R. Potts, *Humanity's Descent: The Consequences of Ecological Instability* (New York: Morrow, 1996).

5. Potts, *Humanity's Descent*, 76–77; R. Cowen, *History of Life*, 3rd ed. (Malden, MA: Blackwell Science, 2000).

6. N. J. Shackleton, J. Backman, H. Zimmerman et al, "Oxygen Isotope Calibration of the Onset of Ice-rafting and History of Glaciation in the North-Atlantic Region," *Nature* 307, no. 5952 (1984): 620–623.

7. Actually, *Homo erectus* left during the Pliocene—*see* section X.

8. B. Senut and M. Pickford, "The dichotomy between African Apes and Humans revisited," *Comptes Rendus Palevol* 3, no. 4 (2004): 265–276.

9. T. R. Disotell, "Human Evolution: Origins of Modern Humans Still Look Recent," *Current Biology* 9, no. 17 (1999): R647–R650. Although most biologists, geneticists, and

anthropologists have accepted this data and the "Out of Africa" theory, a few dissenters remain—*please see* A. G. Thorne and M. H. Wolpoff, "The Multi-regional Evolution of Humans," *Scientific American* 266, no. 4 (1992): 28–33, for an explanation of the rival "Multi-regional hypothesis."

10. T. D. White, B. Asfaw, D. DeGusta, H. Gilbert, G. D. Richards, G. Suwa, and F. C. Howell, "Pleistocene Homo Sapiens from Middle Awash, Ethiopia," *Nature* 423 (2003): 742–747.

11. I. McDougall, F. H. Brown, and J. G. Fleagle, "Stratigraphic Placement and Age of Modern Humans from Kibish, Ethiopia," *Nature* 433 (2005): 733–6.

12. S. McBrearty and A. S. Brooks, "The Revolution That Wasn't: A New Interpretation of the Origin of Modern Human Behavior," *Journal of Human Evolution* 39 (2000): 453–563.

13. Yehai Ke, et al., "African Origin of Modern Humans in East Asia: A Tale of 12,000 Y Chromosomes," *Science* 292, no. 5519 (2001): 1151–1153.

14. R. Potts, *Humanity's Descent: The Consequences of Ecological Instability* (New York: Morrow, 1996).

15. G. F. Miller, *The Mating Mind: How Sexual Choice Shaped the Evolution of Human Nature* (New York: Doubleday, 2000).

16. R. W. Byrne, *The Thinking Ape: Evolutionary Origins of Intelligence* (Oxford and New York: Oxford University Press, 1995).

17. F. B. M. de Waal, *Chimpanzee Politics: Power and Sex among Apes* (New York: Harper and Row, 1982).

18. R. Lee and I. DeVore, eds., *Man the Hunter* (Chicago: Aldine Publishing Company, 1969).

19. M. Sahlins, *Stone Age Economics* (Chicago: Aldine Publishing Company, 1972).

20. P. Tierney, *Darkness in El Dorado: How Scientists and Journalists Devastated the Amazon* (New York: W. W. Norton & Company, Inc., 2000).

21. E. Wilmsen, *Land Filled With Flies: A Political Economy of the Kalahari* (Chicago: University of Chicago Press, 1989): 271.

22. R. Dart, "The Predatory Transition from Ape to Man," *International Anthropological and Linguistic Review* 1 (1953): 201–217.

23. D. C. Johanson and B. Edgar, *From Lucy to Language* (New York: Simon and Schuster, 1996): 91.

24. J. Goodall, The Chimpanzees of Gonde (Cambridge, MA MA: Harvard University Press, 1986).

25. D. Collins and J. Onians, "The Origins of Art," *Art History,* 1 (1978): 1–25.

26. T. Carlyle, *Sartor Resartus,* M. Engel and R. L. Tarr (eds.). (Berkeley: University of California, Press, 2000 [1834]) p. 31.

27. S. Mithen, *Prehistory of the Mind* (London: Routledge, 1996)

28. P. S. Martin and R. G. Klein, eds., *Quaternary Extinctions: A Prehistoric Revolution* (Tucson: The University of Arizona Press, 1984).

29. C. Gamble, *The Palaeolithic Societies of Europe* (Cambridge: Cambridge University Press, 1999); T. D. Price and Gary M. Feinman, *Images of the Past* (Mountain View, CA. and London: Mayfield, 1997).

30. Although a new dating technique puts back the initial occupation at Lake Mungo to 60,000 years ago, *see* A. Thorne, R. Grun, G. Mortimer, N. A. Spooner, J. J. Simpson,

M. McCulloch, L. Taylor, D. Curnoe, "Australia's Oldest Human Remains: Age of the Lake Mungo 3 Skeleton," *Journal of Human Evolution* 36, no. 6 (1999): 591–612.

CHAPTER TWO

1. D. Rindos, *The Origins of Agriculture: An Evolutionary Perspective* (London: Academic Press, 1984).

2. J. Diamond, *Guns, Germs, and Steel,* (New York: W. W. Norton & Company, Inc., 1997 (1999): 114.

3. P. Richerson, R. Boyd, and R. Bettinger, "Was Agriculture Impossible during the Pleistocene but Mandatory during the Holocene? A Climate Change Hypothesis," *American Antiquity* 66, no. 3 (2001): 387–411.

4. Richerson, et al., "Was Agriculture Impossible during the Pleistocene but Mandatory during the Holocene?"

5. M. N. Cohen, *The Food Crisis in Prehistory: Overpopulation and the Origins of Agriculture* (New Haven, CT: Yale University Press, 1977).

6. J. Harlan, "Distributions of Agricultural Origins: A Global Perspective," in A. B. Damania, J. Valkoun, G. Willcox, and C. O. Qualset, eds., *The Origins of Agriculture and Crop Domestication.* Proceedings of the Harlan Symposium, 10–14 May 1997. (Aleppo, Syria: FAO/IPGRI/GRCP/ICARDA, 1998).

7. J. Diamond, "Evolution, Consequences and Future of Plant and Animal Domestication," *Nature* 418, no. 6898 (2002): 700–707.

8. 'Ain Mallaha is the Arabic name of this site, while Eynan is the Hebrew name. I have used' Ain Mallaha because it is the name used in the original excavation reports, as well as in several other standard references.

9. D. Lewis-Williams, "Constructing a Cosmos—Architecture, Power and Domestication at Catalhoyuk," *Journal of Social Archaeology* 4, no. 1 (2004): 28–59.

10. The puzzling uselessness of many of the first domesticates has tempted at least one archaeologist to characterize the first domesticates as luxury goods. Brian Hayden believes that people domesticated plants and animals to provide elaborate feasts for their neighbors—not simply in order to eat some boring staples. *See* his article: B. Hayden, "Nimrods, Piscators, Pluckers and Planters: The Emergence of Food Production," *Journal of Anthropological Archaeology* 9 (1990): 31–69; B. Hayden, "A New Overview of Domestication," in D. T. Price and A. B. Gebauer, eds., *Last Hunters, First Farmers: New Perspectives on the Prehistoric Transition to Agriculture* (Santa Fe, NM: School of American Research Press, 1995): 273–299, B. Hayden, "Were Luxury Foods the First Domesticates? Ethnoarchaeological Perspectives from Southeast Asia," *World Archaeology* 34, no. 3 (2003), 458.

11. B. D. Smith, "The Initial Domestication of Cucurbita Pepo in the Americas 10,000 Years Ago," *Science* 276, no. 5314 (1997): 932–934.

12. D. R. Piperno and K. V. Flannery, "The Earliest Archaeological Maize (*Zea mays L.*) from Highland Mexico: New Accelerator Mass Spectrometry Dates and Their Implications," *Proceedings of the National Academy of Sciences of the United States of America* 98, no. 4 (2001): 2101–2103.

13. D. R. Piperno and D. M. Pearsall, *The Origins of Agriculture in the Lowland Neotropics,* (San Diego and London: Academic Press, 1998).

14. J. Harlan, "Wild-Grass Seed Harvesting in the Sahara and Sub-Sahara of Africa," in D. R. Harris and G. C. Hillman, eds., *Foraging and Farming: The Evolution of Plant Exploitation* (London: Unwin Hyman, 1989): 79–98, 83–84.

15. P. Bellwood and C. Renfrew, *Examining the Language/Farming Dispersal Hypothesis* (Cambridge: McDonald Institute for Archaeological Research, 2002); C. Renfrew, *Archaeology and Language: The Puzzle of Indo-European Origin,* (London: Cambridge University Press, 1987); C. Renfrew, "Language Families and the Spread of Farming," in D. R. Harris, ed., *The Origins and Spread of Agriculture and Pastoralism in Eurasia* (Washington, D.C.: Smithsonian Institution Press, 1996): 70–92; L. L. Cavalli-Sforza, "The Spread of Agriculture and Nomadic Pastoralism: Insights from Genetics, Linguistics, and Archaeology," in D. R. Harris, ed., *The Origins and Spread of Agriculture and Pastoralism in Eurasia* (Washington, D.C.: Smithsonian Institution Press, 1996): 51–69.

16. L. H. Keely, *War Before Civilization* (New York: Oxford University Press, 1996).

17. A. J. Ammerman and L. L. Cavalli-Sforza, *The Neolithic Transition and the Genetics of Populations in Europe* (Princeton, NJ: Princeton University Press, 1984).

18. C. C. Lamberg-Karlovsky, "The Indo-Iranians," *Current Anthropology* 43, no. 1 (2002): 63–88.

19. J. P. Mallory, *In Search of the Indo-Europeans: Language, Archaeology, and Myth* (London: Thames & Hudson, 1991).

20. J. Diamond, *Guns, Germs, and Steel* (New York: W. W. Norton & Company, Inc., 1997 (1999): 384.

21. Diamond, *Guns, Germs, and Steel*, 394.

22. R. Porter, *The Greatest Benefit to Mankind: A Medical History of Humanity* (New York: W. W. Norton & Company, Inc., 1999).

23. P. Smith, O. Bar-Yosef, and A. Sillen, "Archaeological and Skeletal Evidence for Dietary Change During the Late Plesitocenel/Early Holocene in the Levant," in M. N. A. Cohen, George J., eds., *Paleopathology at the Origins of Agriculture* (Orlando and London: Academic Press, 1984): 101–136.

24. V. Eshed, A. Gopher, T. B. Gage, and I. Hershkovitz, "Has the Transition to Agriculture Reshaped the Demographic Structure of Prehistoric Populations? Evidence from the Levant," *American Journal of Physical Anthropology* 124 (2004): 315–329.

25. L. Garrett, *The Coming Plague: Newly Emerging Diseases in a World Out of Balance* (New York and London: Penguin, 1994), 235.

26. D. W. Sellen and R. Mace, "Fertility and Mode of Subsistence: A Phylogenetic Analysis," *Current Anthropology* 38, no. 5 (1997): 878–889.

27. V. Eshed, A. Gopher, T. B. Gage, and I. Hershkovitz, "Has the Transition to Agriculture Reshaped the Demographic Structure of Prehistoric Populations? Evidence from the Levant," *American Journal of Physical Anthropology* 124 (2004): 315–329.

28. R. Pennington, "Hunter-gatherer Demography," in C. Panter-Brick, R. H. Layton, and P. Rowley-Conwy, eds., *Hunter-gatherers: An Interdisciplinary Perspective* (Cambridge: Cambridge University Press, 2001): 170–204.

29. D. E. Yen, "The Domestication of Environment," in D. R. Harris and G. C. Hillman, ed., *Foraging and Farming: The Evolution of Plant Exploitation* (London: Unwin Hyman, 1989): 55–75.

30. R. Jones and B. Meehan, 1989. "Plant Foods of the Gidjingali: Ethnographic and Archaeological Perspectives from Northern Australia on Tuber and Seed Exploitation," in

D. R. Harris and G. C. Hillman (eds.), *Foraging and Farming: The Evolution of Plant Exploitation* (London: Unwin Hyman): 120–135.

31. M. A. Baumhoff, "The Carrying Capacity of Hunter-Gatherers," in S. T. Koyama and David Hurst, eds., *Affluent Foragers: Pacific Coasts East and West. Senri Ethnological Studies* 9 (1981): 77–87.

32. M. Zvelebil and P. Rowley-Conwy, "Foragers and Farmers in Atlantic Europe," in M. Zvelebil, ed., *Hunters in Transition: Mesolithic Societies of Temperate Eurasia and Their Transition to Farming* (Cambridge: Cambridge University Press, 1986): 67–93.

33. I. Kuijt and N. Goring-Morris, "Foraging, Farming, and Social Complexity in the Pre-Pottery Neolithic of the Southern Levant: A Review and Synthesis," *Journal of World Prehistory* 16, no. 4 (2002): 361–440.

34. K. V. Flannery, "The Origins of the Village as a Settlement Type in Mesoamerica and the Near East: A Comparative Study," in P. Ucko, R. Tringham, and G. W. Dimbleby, eds., *Man, Settlement, and Urbanism* (London: Duckworth Publishers, 1972): 23–53.

35. B. F. Byrd, "Public and Private, Domestic and Corporate: The Emergence of the Southwest Asian Village," *American Antiquity* 59 (1994): 639–666; B. F. Byrd, "Households in Transition: Neolithic Social Organization within Southwest Asia," in I. Kuijt, ed., *Life in Neolithic Farming Communities: Social Organization, Identity, and Differentiation* (New York: Springer, 2000): 63–98.

36. K. V. Flannery, "The Origins of the Village Revisited: From Nuclear to Extended Households," *American Antiquity* 67, no. 3 (2002): 417–433.

37. Flannery, "The Origins of the Village Revisited," 417–433.

38. Flannery, "The Origins of the Village Revisited," 417–433.

39. R. Haaland, "Sedentism, Cultivation and Plant Domestication in the Holocene Middle Nile Region," *Journal of Field Archaeology* 22, no. 2 (Summer, 1995): 157–174.

40. V. Eshed, A. Gopher, T. B. Gage, and I. Hershkovitz, "Has the Transition to Agriculture Reshaped the Demographic Structure of Prehistoric Populations? New Evidence from the Levant," *American Journal of Physical Anthropology* 124 (2004): 315–329.

41. T. Molleson, "The Eloquent Bones of Abu-Hureyra," *Scientific American* 271, no. 2 (1994): 70–75.

42. T. L. Jones, "Mortars, Pestles, and Division of Labor in Prehistoric California: A View from Big Sur," *American Antiquity* 61, no. 2 (1996): 243–264.

43. I. Hodder, "Women and Men at Çatalhöyük," *Scientific American* 290, no. 1 (2004): 76–83.

44. L. Garrett, *The Coming Plague: Newly Emerging Diseases in a World Out of Balance* (New York and London: Penguin, 1994), 66.

45. V. G. Childe, *Man Makes Himself* (London, 1956), 66.

CHAPTER THREE

1. C. K. Maisels, *Early Civilizations of the Old World: The Formative Histories of Egypt, the Levant, Mesopotamia, India, and China* (London and New York: Routledge, 1999); N. Yoffee, "Too Many Chiefs? (or Safe Texts for the '90s)," in S. Andrew (ed.), *Archaeological Theory: Who Sets the Agenda?* (Cambridge: Cambridge University Press, 1993): 60–78.

2. K. V. Flannery, "Childe the Evolutionist," in D. R. Harris, ed., *The Archaeology of V. Gordon Childe: Contemporary Perspectives* (Chicago: University Of Chicago Press, 1994): 104–105.

3. M. Sahlins, *Stone Age Economics* (Chicago: Aldine Publishing Company, 1972).

4. D. Malo, *Hawaiian Antiquities* (Honolulu, HI: Bernice P. Bishop Museum Special Publications, 1951).

5. D. Oates and J. Oates, "Early Irrigation Agriculture in Mesopotamia," in K. E. Wilson, ed., *Problems in Economic and Social Archaeology* (London, 1976): 109–135.

6. G. J. Stein, "Economy, Ritual, and Power in 'Ubaid Mesopotamia," in G. J. Stein, ed., *Chiefdoms and Early States in the Near East: The Organizational Dynamics of Complexity. Monographs in World Archaeology* 18, (Madison, WI: 1994): 35–46.

7. F. Hole, "Environmental Instabilities and Urban Origins," in G. J. Stein and M. S. Rothman, eds., *Chiefdoms and Early States in the Near East: The Organizational Dynamics of Complexity. Monographs in World Archaeology* 18, (Madison, WI: 1994): 121–151.

8. B. Fagan, *People of the Earth* (Upper Saddle River, NJ: Prentice Hall, 1998).

9. J. Marcus and K. V. Flannery, *Zapotec Civilization: How Urban Society Evolved in Mexico's Oaxaca Valley, New Aspects of Antiquity* (London: Thames & Hudson, 1996).

10. Marcus and Flannery, *Zapotec Civilization*, 106–107.

11. M. Coe and R. Koontz, *Mexico: From the Olmecs to the Aztecs* (London and New York: Thames & Hudson, 2002).

12. This definition owes an obvious debt to the work of Henry Wright and Gregory Johnson; *see* especially H. Wright and G. Johnson, "Population, Exchange, and Early State Formation in Southwestern Iran," *American Anthropologist* 77 (1975): 267–289.

13. J. Marcus and K. V. Flannery, *Zapotec Civilization*, 156.

14. V. G. Childe, *Man Makes Himself* (London: Spokesman Press, 1956), 134.

15. K. Wittfogel, *Oriental Despotism: A Comparative Study of Total Power* (New Haven, CT, CT: Yale University Press, 1957).

16. R. M. Adams, "Historic Patterns of Mesopotamian Irrigation Agriculture," in M. Gibson, ed., *Irrigation's Impact on Society* (Tucson, AZ: University of Arizona Press, 1974): 1–6.

17. R. M. Adams, *The Evolution of Urban Society: Early Mesopotamia and Prehispanic Mexico* (Chicago: Aldine Publishing Company, 1966).

18. B. R. Foster, *The Epic of Gilgamesh* (New York: W. W. Norton & Company, Inc., 2001).

19. The US Census bureau defines an urban center as any settlement with over 2,500 people. Comparison with France is even more apt because the French government defines an urban center as any settlement with a population greater than 2,000. Only 74 percent of France is urban, meaning that using the same definitions, modern France is less urban than ancient Sumer.

20. F. Hole, "Environmental Instabilities and Urban Origins," in G. J. Stein and M. S. Rothman, eds., *Chiefdoms and Early States in the Near East: The Organizational Dynamics of Complexity. Monographs in World Archaeology* 18, Madison 1994: 121–151.

21. G. Algaze, "Initial Social Complexity in Southwestern Asia—The Mesopotamian Advantage," *Current Anthropology* 42, no. 2 (2001): 199–233.

22. R. J. Wenke, *Patterns in Prehistory: Humankind's First Three Million Years*, (Oxford: Oxford University Press, 1999), 440.

23. B. J. Kemp, *Ancient Egypt: Anatomy of a Civilization* (London and New York: Routledge, 1989), 38

24. B. Fagan, *People of the Earth* (Upper Saddle River, NY: Prentice Hall, 1998).

25. B. J. Kemp, *Ancient Egypt: Anatomy of a Civilization* (London and New York: Routledge, 1989).

26. R. J. Wenke, *Patterns in Prehistory: Humankind's First Three Million Years* (Oxford: Oxford University Press, 1999), 452.

27. I. E. S. Edwards, *The Pyramids of Egypt* (New York: Penguin, 1985).

28. M. E. Moseley, *The Incas and Their Ancestors: The Archeology of Peru*, (London: Thames & Hudson, 1992), 7.

29. Moseley, *The Incas and Their Ancestors*, 8.

30. R. L. Burger, *Chavín and the Origins of Andean Civilization* (London: Thames & Hudson, 1992).

31. Burger, *Chavin and the Origins of Andean Civilization*, 220.

32. A. von Hagen and Craig Morris, *The Cities of the Ancient Andes* (London: Thames & Hudson, 1998).

33. J. Diamond, *Guns, Germs, and Steel* (New York: W. W. Norton & Company, Inc., 1997 [2003]), 281.

CHAPTER FOUR

1. Miriam Lichtheim, *Ancient Egyptian Literature: A Book of Readings. Volume 1: The Old and Middle Kingdoms* (Berkeley, CA: University of California Press, 1972), 185–192.

2. Gary Urton, "From Knots to Narratives: Reconstructing the Art of Historical Record-Keeping in the Andes from Spanish Transcriptions of lnka Khipus," *Ethnohistory* 45, no. 3 (1998): 409–438.

3. Denise Schmandt-Besserat, *How Writing Came About* (Austin, TX: University of Texas Press, 1996), 123.

4. M. del Carmen Rodríguez Martínez, P. Ortíz Ceballos, M. D. Coe, R. A. Diehl, S. D. Houston, K. A. Taube, A. D. Caldéron, "Oldest Writing in the New World," *Science* 313, no. 5793 (2006): 1610–1614.

5. Mary E. D. Pohl, Kevin O. Pope, and Christopher von Nagy, "Olmec Origins of Mesoamerican Writing," *Science* 298, no. 5600 (2002), 1986.

6. Javier Urcid Serrano, *Zapotec Hieroglyphic Writing* (Washington D.C.: Dumbarton Oaks Research Library and Collection, 2001), 409.

7. Nicholas Postgate, Tao Wang, and Toby Wilkinson, "The Evidence for Early Writing: Utilitarian or Ceremonial?" *Antiquity* 69 (1995): 459–480.

8. Anne Underhill, "Craft Production and Social Change in Northern China," *Fundamental Issues in Archaeology* (New York: Kluwer Academic/Plenum Publishers, 2002), 37–38.

9. Kwang-Chih Chang, *Shang Civilization* (New Haven, CT: Yale University Press, 1980), 247-8.

10. Urton, "From Knots to Narratives," 431.

11. Susan Pollock, *Ancient Mesopotamia: The Eden That Never Was* (Cambridge: Cambridge University Press, 1999), 98.

12. Moses Finlay, "Mycenaean Palace Archives and Economic History" in M. Finlay, ed., *Economy and Society in Ancient Greece* (London: Chatto and Windus, 1981), 206.

13. Brian Fagan, *People of the Earth* (New York: Longman, An imprint of Addison Wesley Longman, Inc., 1998).

14. Sofia Voutsaki, "Economic Control, Power and Prestige in the Mycenaean World: The Archaeological Evidence," in J. Killen and S. Voutsaki, eds., *Economy and Politics in the Mycenaean Palace States* (Cambridge: Cambridge Philological Society, 2001): 195–213.

15. Ross Hassig, *War and Society in Ancient Mesoamerica* (Berkeley, CA: University of California Press, 1992), 50.

16. François Lenormant and E. Chevallier, *A Manual of the Ancient History of the East: To the Commencement of the Median Wars* (London: Asher and Co., 1869), 417–8.

17. Gwendolyn Leick, *Mesopotamia: The Invention of the City* (London and New York: Penguin, 2001), 86.

18. Thorkild Jacobsen quoted in Karl Wittfogel, *Oriental Despotism; a Comparative Study of Total Power* (New Haven, CT: Yale University Press, 1978), 267.

19. T. Jacobsen, *Early Political Development in Mesopotamia, Towards the Image of Tammuz* (Cambridge: Cambridge University Press, 1970), 32–56.

20. Marc van de Mieroop, *The Ancient Mesopotamian City* (Cambridge: Cambridge University Press, 1999), 139.

21. A.K. Grayson, *Assyrian Rulers of the Third and Second Millennium BC (to 1115 BC)* (RIMA 1) (Toronto: University of Toronto Press, 1987).

22. This account of the caste system comes from Louis Dumont, *Homo Hierarchicus: The Caste System and Its Implications*, translated from the French by Mark Sainsbury, Louis Dumont, and Basia Gulati (Chicago: University of Chicago Press, 1980).

23. Charles Keith Maisels, *Early Civilizations of the Old World: The Formative Histories of Egypt, the Levant, Mesopotamia, India, and China* (London and New York: Routledge, 1999), 236.

24. Jonathan Mark Kenoyer, *Ancient Cities of the Indus Valley Civilization* (Oxford: Oxford University Press, 1998), 83–84.

25. Douglas Price and Gary Feinman, *Images of the Past* (Madison, WI: University of Wisconsin Press, 1997), 405.

26. Statistics are from the 2000 US Census (www.census.gov).

CHAPTER FIVE

1. S. A. M. Adshead, *Central Asia in World History* (London: Macmillan Press, 1993).

2. W. McNeill, "The Rise of the West after Twenty-five Years," *Journal of World History* 1 (1990): 1–22. Other world historians and archaeologists have also elaborated these views, *see* especially A. G. Frank, "Bronze Age World Systems?" *Current Anthropology* 34 (1993): 383–429; P. L. Kohl, "The Use and Abuse of World Systems Theory: The Case of the 'Pristine' West Asian State," in C. C. Lamberg-Karlosvsky, ed., *Archaeological Thought in America* (Cambridge: Cambridge University Press, 1989): 218–240; and A. Sherratt, "Material Resources, Capital, and Power: The Coevolution of Society and Culture," in G. Feinman and L. Nicholas, eds., *Archaeological Perspectives on Political Economies* (Salt Lake City, UT: University of Utah Press, 2004): 79–103.

3. A. Sherratt, "Gordon Childe: Right or Wrong?" *Economy and Society in Prehistoric Europe: Changing Perspectives* (Edinburgh: Edinburgh University Press, 1995 [1997]): 490–505.

4. P. L. Kohl, "Archaeological Transformations: Crossing the Pastoral/Agricultural Bridge," *Iranica Antiqua* 38 (2002): 151–190.

5. A. Sherratt, "The Secondary Exploitation of Animals in the Old World," *Economy and Society in Prehistoric Europe*, (Edinburgh: Edinburgh University Press, 1997): 199–228; A. Sherratt, "Plough and Pastoralism: Aspects of the Secondary Products Revolution," *Economy and Society in Prehistoric Europe*, (Edinburgh: Edinburgh University Press, 1997 [1981]): 158–198; A. Sherratt, "Wool, Wheels and Ploughmarks: Local Developments or Outside Introductions in Neolithic Europe?" *Economy and Society in Prehistoric Europe*, (Edinburgh: Edinburgh University Press, 1997 [1987]): 229–241.

6. Farmers began milking the occasional sheep, goat, or cow soon after they were domesticated. However, a focus on widespread dairying appeared in the fourth millennium BCE, at the same time as the state. *See* A. Sherratt, "The Secondary Exploitation of Animals in the Old World," *Economy and Society in Prehistoric Europe* (Edinburgh: Edinburgh University Press, 1997): 199–228; A. Sherratt, "Material Resources, Capital, and Power: The Coevolution of Society and Culture," in G. Feinman and L. Nicholas, eds., *Archaeological Perspectives on Political Economies* (Salt Lake City: UT: University of Utah Press, 2004): 79–103.

7. These are technically ards, a type of simple plough.

8. A. Sherratt, "Cups That Cheered: The Introduction of Alcohol to Prehistoric Europe," *Economy and Society in Prehistoric Europe* (Edinburgh: Edinburgh University Press, 1997): 376–402.

9. Herodotus thought the Scythians had discovered an alternative to bathing! A. D. Sélincourt, *Herodotus, The Histories*, rev. A. R. Burn (New York: Penguin, 1972), 295.

10. A. Sherratt, "Cups That Cheered: The Introduction of Alcohol to Prehistoric Europe," *Economy and Society in Prehistoric Europe* (Edinburgh: Edinburgh University Press, 1997): 376–402.

11. B. Fagan, *Ancient North America: The Archaeology of a Continent* (London: Thames & Hudson, 1995).

12. A. Sherratt, "Plough and Pastoralism: Aspects of the Secondary Products Revolution," *Economy and Society in Prehistoric Europe* (Edinburgh: Edinburgh University Press, 1997 [1981]): 158–198; 195–196.

13. S. Jagchid and V. J. Symons, *Peace, War and Trade along the Great Wall: Nomadic-Chinese Interaction through Two Millennia* (Bloomington: Indiana University Press, 1989), 193, note 5.

14. N. Di Cosmo, *Ancient China and Its Enemies: The Rise of Nomadic Power in East Asian History* (Cambridge: Cambridge University Press, 2002).

15. M. Frachetti, "Bronze Age Exploitation and Political Dynamics of the Eastern Eurasian Steppe Zone," *Ancient Interactions: East and West in Eurasia* K. Boyle, C. Renfrew and M. Levine (eds.) (Cambridge: McDonald Institute for Archaeological Research, 2002): 161–170.

16. Several Chinese archaeologists dispute this and argue that the Chinese independently invented bronze metallurgy. However, bronze in the Tarim basin is earlier than any bronze found in East China, supporting this scenario. Z. An, "Cultural Complexes of the Bronze Age in the Tarim Basin and Surrounding Areas," *The Bronze Age and Early Iron Age Peoples of Eastern Central Asia* V. Mair. (Washington, DC: Institute for the Study of Man, 1998): 45–62.

17. A. Khazanov, *Nomads and the Outside World* (Madison, WI: University of Wisconsin Press, 1994).

18. B. W. Young and M. L. Fowler, *Cahokia: The Great Native American Metropolis* (Urbana, IL: University of Illinois, 2000).

19. Young and Fowler, *Cahokia*.

20. C. Calloway, *One Vast Winter Count: The Native American West Before Lewis and Clark* (Lincoln, NE: University of Nebraska Press, 2003).

21. J. A. Brown, "The Archaeology of Ancient Religion in the Eastern Woodlands," *Annual Review of Anthropology* 26 (1997): 465–485.

22. B. Fagan, *Ancient North America: The Archaeology of a Continent* (London: Thames & Hudson, 1995).

23. C. Calloway, *One Vast Winter Count: The Native American West Before Lewis and Clark* (Lincoln, NE: University of Nebraska Press, 2003).

24. B. W. Young and M. L. Fowler, *Cahokia: The Great Native American Metropolis* (Urbana, IL: University of Illinois, 2000).

25. N. Sandars, *The Sea Peoples: Warriors of the Ancient Mediterranean 1250–1150 BC* (London: Thames & Hudson, 1978). Translation is mine.

26. R. Drews, *The End of the Bronze Age: Changes in Warfare and the Catastrophe ca. 1200 BC* (Princeton: Princeton University Press, 1993).

27. Drews, *The End of the Bronze Age*.

28. N. Sandars, *The Sea Peoples: Warriors of the Ancient Mediterranean 1250–1150 BC* (London: Thames & Hudson, 1978). Translation is mine.

29. But *see* M. Mieroop, *The Ancient Near East* (Oxford: Blackwells, 2004): 182.

FOR FURTHER READING

Brian Fagan's *People of the Earth* (Upper Saddle River, NJ: Prentice Hall, 2003), Robert Wenke's *Patterns in Prehistory* (New York: Oxford University Press, 1999), and Douglas Price and Gary Feinman's *Images of the Past* (New York: McGraw-Hill, 2004) are well-illustrated introductions to early history. Jared Diamond's *Guns, Germs and Steel* (New York: Norton, 1997) discusses the origins of agriculture and its consequences for later human societies. Marshall Sahlins's *Stone Age Economics* (New York: Aldine, 1972) is a classic account of how foragers, villagers, and people in urban societies make their living.

For students interested in learning more about human evolution, Roger Lewin's *Human Evolution: An Illustrated Introduction* (Malden, MA: Blackwell, 2004) is a good place to start. Steven Mithen's *Prehistory of the Mind* (London: Thames and Hudson, 1996) is an exciting account of prehistory, focusing on how human intelligence first appeared. Scientific reports about the genetic and skeletal evidence for human evolution appear frequently in science journals such as *Science* and *Nature*, and *Science News* and *Nature News* provide information that a nonscientist can understand.

Two recent books explore the evidence for the transition to agriculture: Steven Mithen's playful *After the Ice: A Global Human History 20,000–5,000 BC* (Cambridge, MA: Harvard University Press, 2004) and Peter Bellwood's *First Farmers: The Origins of Agricultural Societies* (Malden, MA: Blackwell, 2004), which focuses on the spread of languages and people following the emergence of farming. Bruce Smith's *The Emergence of Agriculture* (New York: Scientific American Library, 1998) and the essays in T. Douglas Price and Anne Gebauer's (eds.) *Last Hunters, First Farmers: New Perspectives on the Prehistoric Transition to Agriculture* (Santa Fe, NM: Schools of American Research Press, 1995) are also interesting. Two articles provide new details on the origins of the village and gender roles: Kent Flannery's "The Origins of the Village Revisited: From Nuclear to Extended Households, *American Antiquity* 7, no. 3 (2002), pp. 417–433, and Ian Hodder's "Women and Men at Çatalhöyük," *Scientific American* 290, no. 1 (2004), pp. 76–83.

The essays in Gary Feinman and Joyce Marcus, (eds.) *Archaic States* (Santa Fe, NM: Schools of American Research Press, 1998) are a great starting place for why states emerged and what made the world's first cities and states so distinctive. Guillermo Algaze's 2001 article "Initial Social Complexity in Southwestern Asia—The Mesopotamian Advantage," *Current Anthropology* 42, no. 2: 199–233 and J. N. Postgate's *Early Mesopotamia: Society and Economy at the Dawn of History* (New York: Routledge, 1994) look at the first states in Mesopotamia, and Joyce Marcus and Kent Flannery's *Zapotec Civilization: How Urban Society Evolved in Mexico's Oaxaca Valley* (New York: Thames and Hudson, 1996) discusses the origins of the state in Mesoamerica. Michael Moseley's *The Incas and Their Ancestors: The Archaeology of Peru* (New York: Thames and Hudson, 2001) sums up the evidence for complex societies in the Andes; Anne Underhill considers the evidence for early states in China in *Craft Production and Social Change in Northern China* (New York: Kluwer Academic/Plenum Publishers, 2002), and John Baines and Jaromir Malek's *Cultural Atlas of Ancient Egypt* (New York: Facts on File, 2000) has great illustrations and succinct descriptions of ancient Egyptian society. Finally, the essays in Andrew Sherratt's *Economy and Society in Prehistoric Europe* (Edinburgh: Edinburgh University Press, 1997) discuss the secondary products revolution and provide an excellent introduction to the study of connections in early history.

PHOTO CREDITS

◖ INDEX ◗